Auto Body Repair Technology

Seventh Edition

Student Technician's Manual

Paul Uhrina
James E. Duffy
Jonathan Beaty
Terry Sparks

Australia • Brazil • Canada • Mexico • Singapore • United Kingdom • United States

Auto Body Repair Technology—Student Technician's Manual, **7th Edition**
Paul Uhrina/James E. Duffy/Jonathan Beaty/ Terry Sparks

SVP, Higher Education & Skills Product: Erin Joyner

Product Director: Mathew Seeley

Senior Product Manager: Katie McGuire

Product Assistant: Kimberly Klotz

Director, Learning Design: Rebecca von Gillern

Senior Manager, Learning Design: Leigh Hefferon

Senior Learning Designer: Mary Clyne

Senior Marketing Director: Michele McTighe

Senior Marketing Manager: Scott Chrysler

Director, Content Creation: Juliet Steiner

Manager, Content Creation: Alexis Ferraro

Senior Content Manager: Sharon Chambliss

Digital Delivery Lead: Amanda Ryan

Art Director: Jack Pendleton

Text Designer: Erin Griffin

Cover Designer: Felicia Bennett

Cover image(s): Max Anuchkin/ShutterStock.com

For product information and technology assistance, contact us at
**Cengage Customer & Sales Support, 1-800-354-9706
or support.cengage.com.**

For permission to use material from this text or product, submit all requests online at **www.cengage.com/permissions.**

Library of Congress Control Number: 2020942976

Book Only ISBN: 978-0-357-13980-6

Cengage
200 Pier 4 Boulevard
Boston, MA 02210
USA

Cengage is a leading provider of customized learning solutions with employees residing in nearly 40 different countries and sales in more than 125 countries around the world. Find your local representative at **www.cengage.com.**

To learn more about Cengage platforms and services, register or access your online learning solution, or purchase materials for your course, visit **www.cengage.com.**

Contents

Collision Repair:
Introduction and Careers

SHOP ASSIGNMENT
1-1

Name _____ Date _____ Instructor Review _____

Occupations

Explain the job duties of the following occupations associated with the collision repair center. Describe each job location in your shop, if applicable.

1. Shop estimator _____
   _____
   _____

   Location _____

2. Insurance adjuster _____
   _____
   _____

   Location _____

3. Metalworking technician _____
   _____
   _____

   Location _____

4. Refinishing technician _____
   _____
   _____

   Location _____

5. Apprentice _____

_____

_____

Location _____

6. Shop owner _____

_____

_____

Location _____

7. Shop foreman _____

_____

_____

Location _____

8. Parts manager _____

_____

_____

Location _____

9. Office manager _____

_____

_____

Location _____

10. Receptionist _____

_____

_____

Location _____

**SHOP ASSIGNMENT
1-2**

Name _____ Date _____ Instructor Review _____

Associated Occupations

Describe the job duties of these specialized occupations associated with the collision repair field. List the experience or education needed to attain these positions.

1. Insurance adjuster _____
   _____
   _____

2. Vocational/technical instructor _____
   _____
   _____

3. Salvage yard technician _____
   _____
   _____

4. Dealership parts person _____
   _____
   _____

5. Paint company representative _____
   _____
   _____

6. Auto manufacturer representative _____
   _____
   _____

7. Equipment salesperson _____
   _____
   _____

8. Glass installer _____
   _____
   _____

9. Paintless dent repair technician _____
   _____
   _____

Associated Occupations

Name _____ Date _____ Instructor Review _____

Job Identification

Match the following jobs to the appropriate pictures.

1. Paint preparation technician _____

2. Structural technician _____

3. Estimator _____

4. Paintless dent repair technician _____

5. Refinish technician _____

6. Customer service specialist _____

7. Detail technician _____

8. Nonstructural technician _____

1

2

3

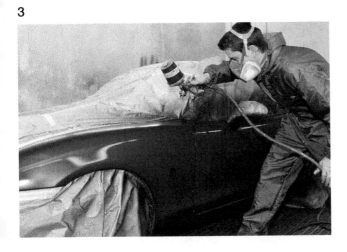

4

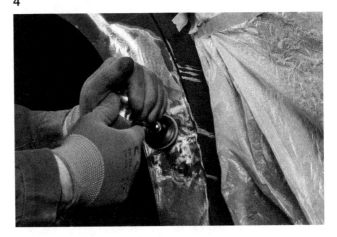

5

6

7

8

Name _____ Date _____

Collision Repair Center Layout

Objective

Upon completion of this activity sheet, the student should be able to recognize and locate the different workstations in any auto body repair shop.

ASE Education Foundation Task Correlation

There are no ASE Education Foundation tasks related to or assessed in this job sheet.

We Support

ASE | **Education Foundation**

Tools and Equipment

Any body shop

Safety Equipment

Safety glasses or goggles

Introduction

Auto body repair shops are very expensive to start up and operate. The shop must be designed to work efficiently with little waste of time, material, and energy. Compare the shop in your school to other shops you may visit. It is a good habit to always wear safety glasses in the shop regardless of the job you are performing.

1. Which chapter and pages explain shop layout? _____

Procedure

2. Draw the layout of your school's shop. In the drawing, indicate the location of the following, labeling them with their respective letters. Upload the drawing using the File Upload button at the bottom of this worksheet when complete.

   a. Fire extinguishers
   b. Lifts
   c. Spray booth(s)
   d. Frame machine
   e. Metalworking stalls
   f. Paint preparation stalls
   g. Cleanup stalls
   h. Air compressor
   i. Air regulators
   j. Tool storage
   k. Locker area
   l. Overhead doors

   m. Emergency exits
   n. Vent fans
   o. Paint mixing area
   p. Paint storage area
   q. Hazardous waste disposal
   r. Spray gun cleaning area
   s. Workbenches
   t. Emergency eye flushing station
   u. First-aid station
   v. Hand-washing area
   w. Vacuum system

3. In your shop layout drawing, use letters to illustrate the flow of a job from start to finish. Use the space provided next to describe the work done at each workstation.

   a. _____

   b. _____

   c. _____

   d. _____

   e. _____

   f. _____

4. Safety is an important issue in a successful body shop. Shops should never be overcrowded with vehicles. Vehicles should be held in a fenced-in area outside the shop when they are not being worked on.

   a. Does your shop have a storage area for vehicles? _____

   b. If so, where is it located? _____

   c. How many vehicles can be stored in this area? _____

   d. How many vehicles can be safely worked on in your shop? _____

   e. How many students work in your school's shop? _____

INSTRUCTOR'S COMMENTS _____

_____

_____

Review Questions

Name _____ Date _____ Instructor Review _____

1. In a collision repair shop, a helper is also called a(n) _____.

2. _____ repairs are usually performed by the helper.

3. Refinishing should only be done in the _____.

4. Insurance adjusters try to make as much money as possible for the collision shop.
   A. True
   B. False

5. Shop estimators are responsible for ordering parts.
   A. True
   B. False

6. Apprentices perform major repairs.
   A. True
   B. False

7. When metalworking, Technician A says that safety glasses are only needed when grinding. Technician B says that safety glasses should be worn all the time. Who is correct?
   A. Technician A
   B. Technician B
   C. Both Technician A and Technician B
   D. Neither Technician A nor Technician B

8. Technician A says that any kind of fire extinguisher is acceptable in a collision center. Technician B says that specific extinguishers must be used. Who is correct?
   A. Technician A
   B. Technician B
   C. Both Technician A and Technician B
   D. Neither Technician A nor Technician B

9. Technician A says that paint must be mixed in an enclosed mixing room. Technician B says that paint can be mixed anywhere in the shop. Who is correct?
   A. Technician A
   B. Technician B
   C. Both Technician A and Technician B
   D. Neither Technician A nor Technician B

10. A vehicle involved in a collision is first brought into the shop, where it is prepared to calculate the cost of repairs. Technician A says this is called making a damage estimate. Technician B says this is also called blueprinting. Who is correct?
    A. Technician A
    B. Technician B
    C. Both Technician A and Technician B
    D. Neither Technician A nor Technician B

Shop Safety

Name _____ Date _____ Instructor Review _____

Personal Safety and Health Protection

Identify and explain the proper use of each item in the illustrations that follow.

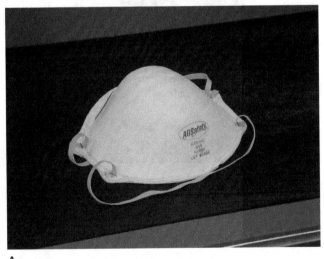

A. _____

    Use _____

B. _____

    Use _____

Courtesy of PPG Industries

C. _____

   Use _____

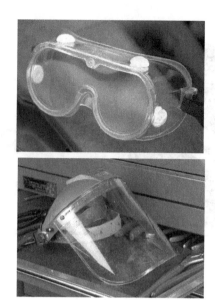

D. _____

   Use _____

E. _____

   Use _____

F. _____

   Use _____

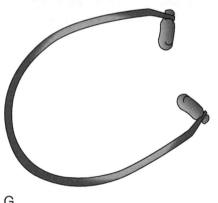

G. _____

   Use _____

Name _____ Date _____ Instructor Review _____

Fire Extinguisher Identification

In the space next to each class of fire extinguisher, list at least three types of fuel material that can be extinguished.

	Class of Fire	Typical Fuel Involved	Type of Extinguisher
Class **A** Fires (green)	**For Ordinary Combustibles** Put out a class A fire by lowering its temperature or by coating the burning combustibles.		Water Foam Multipurpose dry chemical
Class **B** Fires (red)	**For Flammable Liquids** Put out a class B fire by smothering it. Use an extinguisher that gives a blanketing, flame-interrupting effect; cover whole flaming liquid surface.		Foam Carbon dioxide Halogenated agent Standard dry chemical Purple K dry chemical Multipurpose dry chemical
Class **C** Fires (blue)	**For Electrical Equipment** Put out a class C fire by shutting off power as quickly as possible and by always using a nonconducting extinguishing agent to prevent electric shock.		Carbon dioxide Halogenated agent Standard dry chemical Purple K dry chemical Multipurpose dry chemical
Class **D** Fires (yellow)	**For Combustible Metals** Put out a class D fire of metal chips, turnings, or shavings by smothering or coating with a specially designed extinguishing agent.		Dry powder extinguishers and agents only

Name _____ Date _____

Collision Repair Shop Safety

Objective

Upon completion of this activity sheet, the student should be able to recognize, locate, and use safety equipment in any collision center.

ASE Education Foundation Task Correlation

IV.A.2   Identify safety and personal health hazards according to OSHA guidelines and the "Right to Know Law." **(HP-I)**

We Support

ASE | Education Foundation

Tools and Equipment

Any collision center
Flashlight

Safety Equipment

Safety glasses or goggles

Introduction

Knowledge of collision repair safety and shop layout is extremely important because of the high-powered equipment, electronics, flammables, and toxic materials that are used. Collision repair centers are subject to OSHA, EPA, and various local regulatory agencies. If the safety standards of these agencies are not adhered to, serious injury or death can occur, as well as the risk of fines and possible closing of the collision center. This job sheet will help you get familiar with the first-aid, safety equipment, and exit locations in the shop.

Procedure

Task Completed

1. Eye protection is required at all times in most shops to comply with OSHA standards. Eye protection should be worn when working with grinders, sanders, drills, chisels, and broken glass and when mixing paint.   ☐

   a.   What do safety glasses protect against? _____

   b.   Is it OK to weld while wearing safety goggles? Explain. _____

   _____

2. In collision repair centers, emergency exits must be clearly marked. Use of welding equipment with paints, thinners, oily rags, and other combustibles makes fires a very real danger. Doors and walkways must be free of clutter and in good working order. Locate all the emergency and safety warning signs.   ☐

   a.   How many exits are there? Where are they located? _____

   _____

3. Locate the first-aid station. Minor cuts and bruises are a common occurrence in the collision repair center because of all the broken and torn sheet metal. Even minor cuts can become infected very quickly, so a first-aid kit should contain sterile gauze, bandages, scissors, antiseptics, and gloves. Never try to treat major cuts or eye injuries yourself; always get to an emergency room. The first-aid kits are usually in the office or classroom.   ☐

   a.   Where is your shop's first-aid kit located?_____

4. Collision repair centers are a dusty environment. Dust from sanding, particles from grinding, and fumes from painting can all irritate the delicate tissues of the eye. If a piece of dust gets in your eye, you should never rub it; this can further scratch the eyeball, leading to permanent damage. An eye flushing station can gently wash these dust particles from your eye. ☐

    a.   Is your shop equipped with one or more eye flushing stations? Where are they?

        _____

    b.   How does the station work? _____

        _____

5. Locate the locker area. ☐

    a.   Why is it important to store personal items and street clothing in a locker?

        _____

6. Does your shop have showers? ☐

    _____

    a.   Explain how plastic filler and fiberglass dust are washed from the skin.

        _____

7. Locate the workbenches in your shop. ☐

    a.   How many are there? _____

    b.   Are there bench grinders on the benches? _____

    c.   If so, are there shields on the grinders? _____

    d.   What is the purpose of the shields? _____

    e.   If there are shields on the grinders, do you still have to wear your safety glasses while operating the grinders? Explain. _____

        _____

8. Does the bench have a vise? _____ ☐

    a.   How is a vise used safely? _____

    b.   Must safety glasses be worn when using the vise? _____

9. Locate the fire safety cabinets. ☐

    a.   How are they labeled? _____

    b.   What is the color of the cabinets? Explain the reason for each color. _____

        _____

    c.   Where are the cabinets located, and what is stored in them? _____

    d.   Can soiled rags be stored in these cabinets? Why or why not? _____

        _____

    e.   What is spontaneous combustion? _____

    f.   Are these cabinets OSHA approved? _____

    g.   Are these cabinets vented? _____

    h.   Are there OSHA-approved metal cans in your shop for soiled rags? _____

    i.   How many are there, and where are they located? _____

        _____

10. Locate any fire extinguishers. ☐

    a. How many extinguishers did you find, and where are they located? _____

    _____

    b. Locate the inspection date on one extinguisher. What is it? _____

    c. Is there a large letter on the extinguisher? If so, what is it and what does it stand for?

    _____

    d. Explain what type of fire can be extinguished with each of the following extinguisher.

    Class A _____

    Class B _____

    Class C _____

    Class D _____

    e. Once an extinguisher is used, what should you do with it? _____

11. Locate all the extension cords in your shop. ☐

    a. How many cords are there, and what is the condition of them? _____

    _____

    b. Why is it important to roll up and store these cords after they are used?

    _____

12. Is your shop equipped with dust control workstations or fume filtration units? ☐

    _____

    a. Where are the filters located? _____

    b. How often must these filters be changed? _____

    c. Can painting be performed at or near these workstations? Explain. _____

    _____

    d. What kind of respirator must be worn when working at or near these workstations?

    _____

INSTRUCTOR'S COMMENTS _____

_____

_____

Name _____ Date _____

Jack and Jack Stand Safety

Objective

Upon completion of this activity sheet, the student should be able to safely maintain and use jacks and jack stands in any collision center.

ASE Education Foundation Task Correlation

There are no ASE Education Foundation tasks related to or assessed in this job sheet.

We Support

ASE | Education Foundation

Tools and Equipment

Any collision center
Vehicle
Appropriate jacks and jack stands
Flashlight
Tape measure

Safety Equipment

Safety glasses or goggles

Introduction

Jacks of all types are common in the collision center. Though they are critical to shop operations, they are often overlooked and taken for granted. These tools must safely lift and hold tons of weight while technicians work under a vehicle. This job sheet will help you learn to safely maintain and use jacks and jack stands to lift and safely support a vehicle for inspection or repair. If you have any difficulty or doubt about any aspect of this activity, be sure to ask your instructor for help or to answer your questions.

Vehicle Description

Year _____ Make _____ Model _____

Procedure

Task Completed

1. Before starting work, it is very important to adhere to all safety standards by using only properly functioning equipment within the weight limits recommended by the manufacturer. ☐

   a. What type of jack are you using? _____

   b. What is the weight limit of the jack? _____

2. The most commonly used jack in a shop is the service jack. These jacks have small hydraulic rams in them that move up and down by turning the handle of the jack. Because these jacks are hydraulic, the fluid must be checked on a regular basis or slipping can occur, causing serious injury and damage. Locate the nut to check the oil. ☐

   a. Is it full? If not, fill it with the appropriate type and amount of recommended fluid.

   b. What kind of oil is used, and how much oil does the pump hold?

   _____

Task Completed

3. There are two main types of vehicle frames: conventional and unibody. Each has its own procedure for safe lifting. ☐

    a. What is a conventional frame? _____

    b. What is a unibody frame? _____

    c. What type of frame are you working on? _____

Conventional Frame

4. Make sure the vehicle is in park with the emergency brake on. When lifting the front end of a vehicle with a conventional frame, place the jack directly under the main cross member. Making sure the jack is centered, lift the vehicle high enough so that the tires are off the ground. Place a jack stand under the frame where the door meets the fender. ☐

    a. What is the weight limit of this jack stand? _____

5. Repeat the same for the other side. Be extremely careful not to put your hands or legs anywhere under the vehicle until the jack stands are securely in place. **BEFORE** lowering the jack onto the jack stands, have your instructor check the jack stand placement. ☐

6. Remove the jack, move to the rear of the vehicle, and place the jack under the differential. ☐

    a. What does a differential look like? _____

7. Making sure the jack is centered, slowly lift the vehicle until the rear tires are off the ground. While lifting the vehicle, have a partner watch the front jack stands to make sure there is no movement. Before placing the rear jacks, make sure they are extended to the same height as the front jacks. ☐

8. Set the jack stands. **BEFORE** removing the jack, have your instructor check its placement. Remove the jack. ☐

Unibody

9. Because unibody vehicles technically have no frame, lifting procedures for these vehicles are different. You may try lifting from the bottom center of the radiator core support or the front frame rails. ☐

    a. Where is the core support located on a unibody vehicle?

    _____

10. Before placing the jack stands, check the vehicle manual for jack placement. Many times, you can place the jack stands in this same area. ☐

    a. Where did you place the front jack stands? _____

11. Have your instructor check your placement of the jack stands **BEFORE** removing the jack. Once your instructor has approved your work, move the jack to the rear of the vehicle and place the jack under the rear axle or frame rails. ☐

12. Lift the vehicle from the proper location until the rear tires are off the ground. Have your partner watch the front jack stands for movement. Place the jack stands under the axle near the wheels. Have your instructor check the placement of the rear stands **BEFORE** removing the jack. ☐

INSTRUCTOR'S COMMENTS _____

_____

_____

Review Questions

Name _____ Date _____ Instructor Review _____

1. _____ should be worn when using tools that create loud noise.

2. Gasoline should be extinguished with a class _____ fire extinguisher.

3. A fresh air–supplied system is used when _____.

4. Welding goggles can be used when MIG welding.
   A. True
   B. False

5. A dust mask is all that is needed when spraying primer.
   A. True
   B. False

6. Grease and oil can be extinguished with a class B extinguisher.
   A. True
   B. False

7. Technician A says that a full face shield is necessary to MIG weld. Technician B says that welding goggles are sufficient. Who is correct?
   A. Technician A
   B. Technician B
   C. Both Technician A and Technician B
   D. Neither Technician A nor Technician B

8. Technician A says that ear plugs can be worn at all times in the shop. Technician B says that they are only necessary around loud noise. Who is correct?
   A. Technician A
   B. Technician B
   C. Both Technician A and Technician B
   D. Neither Technician A nor Technician B

9. Technician A says that goggles should be worn when handling fluids that can burn your eyes. Technician B says that safety glasses are sufficient. Who is correct?
   A. Technician A
   B. Technician B
   C. Both Technician A and Technician B
   D. Neither Technician A nor Technician B

10. Technician A says that a fresh air–supplied system should be used when sanding fiberglass. Technician B says that it is only needed when painting. Who is correct?
    A. Technician A
    B. Technician B
    C. Both Technician A and Technician B
    D. Neither Technician A nor Technician B

Vehicle Construction Technology

Name _____ Date _____ Instructor Review _____

Vehicle Parts

1. What type of frame does this car have? _____

2. Write the number for each part beside its correct name.

_____ Rear bumper   _____ Fender
_____ Rocker panel   _____ Fuel door/gas lid
_____ Front wheel   _____ Rear door
_____ Sail panel   _____ Front bumper
_____ Front door   _____ Quarter panel
_____ Rear wheel

_____ Front bumper cover   _____ Fog light cover
_____ Hood   _____ Windshield
_____ Lower grille   _____ Grille
_____ Headlight

_____ High stop lamp
_____ Taillights
_____ Deck lid
_____ Rear bumper
_____ Back glass

3. What type of frame does this vehicle have? _____

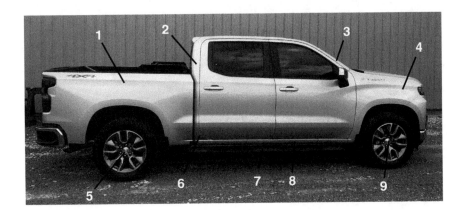

4. Write the number for each part beside its correct name.

_____ Step bar   _____ Right windshield post
_____ Right rear wheel   _____ Right front wheel
_____ Right bedside   _____ Rocker panel
_____ Cab corner   _____ Right fender
_____ Cab

Name _____ Date _____ Instructor Review _____

Vehicle Identification

From memory, identify the types of vehicles and sections of the frame shown in the illustrations. When you have finished, show your answers to your instructor. Then, using your textbook, fill in any you may have missed.

A. _____

B. _____

C. _____

D. _____

E. _____

F. _____

G. _____

H. _____

I. _____

Conventional Frame

A. _____   E. _____

B. _____   F. _____

C. _____   G. _____

D. _____   H. _____

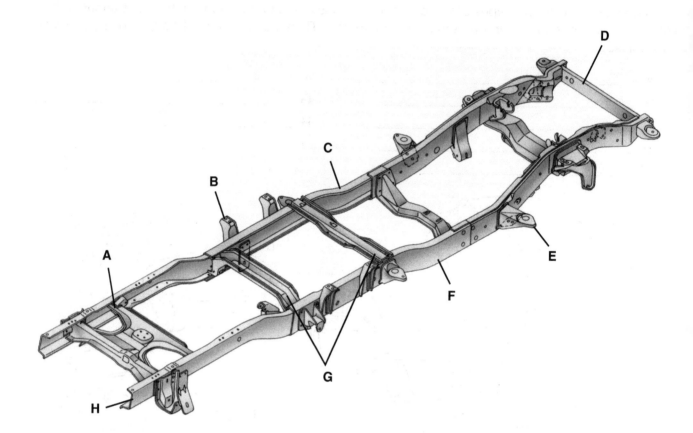

Typical Unibody Construction

A. _____

B. _____

C. _____

D. _____

E. _____

F. _____

G. _____

H. _____

I. _____

J. _____

K. _____

L. _____

M. _____

N. _____

O. _____

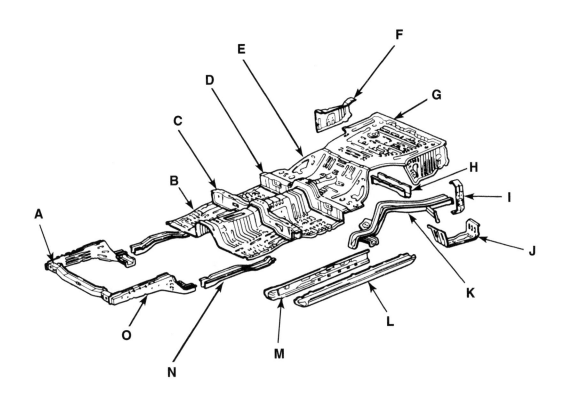

Name _____ Date _____

Older and Newer Vehicle Comparison

Objective

Upon completion of this activity sheet, the student should be able to wash a vehicle with soap and water. Also, the student should be able to identify, locate, and measure the different components of any given older vehicle and then compare the data to a newer vehicle.

ASE Education Foundation Task Correlation

II.B.7   Soap and water wash entire vehicle; complete pre-repair inspection checklist. **(HP-I)**

V.D.1   Identify the type of vehicle construction (space frame, unibody, body-over-frame). **(HP-G)**

V.D.2   Recognize the different damage characteristics of space frame, unibody, and body-over-frame vehicles. **(HP-G)**

We Support

ASE | Education Foundation

Tools and Equipment

Any vehicle 20 years or older
Any modern vehicle
Tape measure
Car wash supplies (wash mitt, soap and water, bucket, tire brush, wheel cleaner or degreaser, chamois)

Safety Equipment

Safety glasses or goggles

Introduction

For many years, American vehicles have been big, heavy gas guzzlers. Until recently, safety features and fuel consumption were not important issues in vehicle design. This exercise will show you the progress made in vehicle design. First, you must clean the vehicle so that you can effectively see the differences and get accurate measurements.

Older Vehicle Description

Note the following data for the specific vehicle you will work on for this job:

Year _____ Make _____ Model _____

VIN _____ Paint code _____

Procedure

Task Completed

1. Before starting work on any vehicle, it is important to review all safety standards pertaining to the job you are performing.   ☐

   a.   Are there any special safety rules that must be followed for this exercise? If so, what are they?

   _____

   _____

2. First, you must wash the exterior of the vehicle so that you can effectively see the differences and get accurate measurements.   ☐

Task Completed

3. Get the car wash supplies listed earlier, and mix a small amount of soap into a clean bucket of water. Rinse any loose dust, dirt, or debris off with the water hose first. ☐

4. Use the appropriate wheel cleaner or degreaser on the wheels and tires to soften up the brake dust. Then use the tire brush to scrub and remove brake dust from the wheels and tires. Then rinse until clean. ☐

5. Start hand washing the vehicle with soap and water and the mitt, starting at the top of the vehicle first. Work your way down the sides of the vehicle so that the last thing you wash is the rocker panels. ☐

  a. Why do we start from the top and work our way down to the lower parts of the vehicle?

  _____

  _____

6. Now dry off the entire vehicle with the chamois. ☐

  a. What will happen if you do not dry the vehicle off? _____

  _____

  b. Now that the vehicle is clean, does it have a metallic paint job?

  _____

  c. How can you tell? _____

7. Get a floor squeegee, and clean up the water around your area into an appropriate floor drain. ☐

8. Depending on the year and model of the vehicle you are inspecting, you may notice that the body of the vehicle is much taller and longer than those of today's vehicles. Gas mileage was not a factor in design, so aerodynamics was not stressed. ☐

  a. What is the length of the vehicle from bumper to bumper?

  _____

  b. What type of bumpers does this vehicle have?

  _____

  c. What are the body panels made of?

  _____

9. Many of today's full-size vehicles would have been considered mid-size in the 1950s, 1960s, and 1970s. ☐

  a. What type of frame does this vehicle have?

  _____

  b. Would this make it stronger than today's unibody vehicles? Explain your answer.

  _____

  _____

  _____

10. Many early vehicles had no safety features. No seat belts and metal dashboards meant ☐
serious head injuries. As vehicles got smaller, more safety features were added. Seat
belts, air bags, padded dashboards, crush zones, and so on, all add to the safety of
the passengers and save many lives.

   a.   Does this vehicle have seat belts? _____

   b.   Are there headrests on the seats? If so, explain how this helps avoid certain injuries.

   _____

   _____

   _____

   c.   Does this vehicle have air bags? _____

   d.   Is there any protection if there is a side impact? If so, explain.

   _____

   _____

   _____

Newer Vehicle Description

Note the following data for the specific vehicle you will work on for this job:

Year _____ Make _____

Model _____ Paint code _____

Location of paint code _____

Procedure

11. Before starting work on any vehicle, it is important to review all safety standards ☐
pertaining to the job you are performing.

12. Are there any special safety rules that must be followed for this exercise? If so, what ☐
are they?

_____

_____

_____

13. First, you must wash the exterior of the vehicle so that you can effectively see the ☐
differences and get accurate measurements.

14. Get the car wash supplies listed earlier, and mix a small amount of soap into a clean ☐
bucket of water. Rinse any loose dust, dirt, or debris off with the water hose first.

15. Use the appropriate wheel cleaner or degreaser on the wheels and tires to soften up the ☐
brake dust. Then use the tire brush to scrub and remove brake dust from the wheels
and tires. Then rinse until clean.

16. Start hand washing the vehicle with soap and water and the mitt, starting at the top of the vehicle first. Work your way down the sides of the vehicle so that the last thing you wash is the rocker panels. ☐

   a. Why do we start from the top and work our way down to the lower parts of the vehicle?

   _____

   _____

   _____

17. Now dry off the entire vehicle with the chamois. ☐

   a. What will happen if you do not dry the vehicle off? _____

   b. Does this vehicle have a solid color, metallic paint, or a tri-coat finish paint job? (**NOTE:** There is a difference in metallic and tri-coat. Be sure you check carefully.)

   _____

   c. How can you tell?

   _____

18. Get a floor squeegee, and clean up the water around your area into an appropriate floor drain. ☐

19. Today's vehicles are rounded and aerodynamic. This cuts down on wind resistance to help them get better gas mileage. You will also notice the difference in size compared to the vehicle in the previous exercise. ☐

   a. What is the length of the vehicle from bumper to bumper?

   _____

   b. What type of bumpers does this vehicle have?

   _____

   How did you determine this?

   _____

   c. What are the body panels made of?

   _____

20. Many of today's full-size vehicles would have been considered mid-size in the 1950s, 1960s, and 1970s. ☐

   a. What type of frame does this vehicle have?

   _____

   b. If a unibody, would this make it stronger than the conventional body-over-frame design?

   _____

   Explain your answer.

   _____

   _____

   _____

21. Today's vehicles are smaller than older ones but safer because of many new safety ☐
features. Seat belts, air bags, padded dashboards, crush zones, and so on, all add to
the safety of the passengers and save many lives.

    a.   What kind of restraint system does this vehicle have?

_____

    b.   Explain how shoulder harnesses help avoid serious head injuries.

_____

_____

    c.   Does this vehicle have more than one air bag? If so, where are they located?

_____

_____

    d.   Explain how headrests help in avoiding injury.

_____

_____

    e.   Is there any protection if there is a side impact? Explain. _____

_____

_____

INSTRUCTOR'S COMMENTS _____

_____

_____

Review Questions

Name _____ Date _____ Instructor Review _____

1. Vehicles that are constructed with the body and frame as one are considered to have _____ frames.

2. The thickness of the metal is measured by _____.

3. Some manufacturers do not recommend straightening UHSS frames.
   A. True
   B. False

4. Crush zones are added to the vehicle after the repairs.
   A. True
   B. False

5. Most of today's vehicles have a clearcoat finish.
   A. True
   B. False

6. Sedans are also called hardtops.
   A. True
   B. False

7. Technician A says that a front cross member is found on a conventional frame. Technician B says that a front cross member is found on a unibody frame. Who is correct?
   A. Technician A
   B. Technician B
   C. Both Technician A and Technician B
   D. Neither Technician A nor Technician B

8. Technician A says that all late model vehicles have air bags in the doors to protect passengers during side impacts. Technician B says that no late model vehicles have air bags in the doors. Who is correct?
   A. Technician A
   B. Technician B
   C. Both Technician A and Technician B
   D. Neither Technician A nor Technician B

9. Technician A says that no late model vehicles have the body-over-frame construction. Technician B says that some full-size vehicles still have a conventional frame. Who is correct?
   A. Technician A
   B. Technician B
   C. Both Technician A and Technician B
   D. Neither Technician A nor Technician B

10. Technician A says that the gauge of the metal does not change the weight of the panel. Technician B says that a difference in gauge will change the weight. Who is correct?
    A. Technician A
    B. Technician B
    C. Both Technician A and Technician B
    D. Neither Technician A nor Technician B

Name _____ Date _____ Instructor Review _____

Abbreviations

From memory, write out what these abbreviations stand for. When you have finished, use the appropriate number to show the part location on the illustrations that follow.

1. A/C _____
2. ALT _____
3. BATT _____
4. CH _____
5. COMP _____
6. CR/CONT _____
7. CV _____
8. DIFF'L _____
9. DOHC _____
10. DRL _____
11. EFI _____
12. EXH _____
13. FRT _____
14. FWD _____
15. H/LAMP _____
16. HVAC _____

17. PNL _____
18. OTR _____
19. RAD _____
20. REG _____
21. RWD _____
22. SIPS _____
23. SPEEDO _____
24. SRS _____
25. SIR _____
26. SUSP _____
27. TRANS _____
28. VIN _____
29. WHLHSE _____
30. W/S _____
31. W/STRIP _____
32. XMBR _____

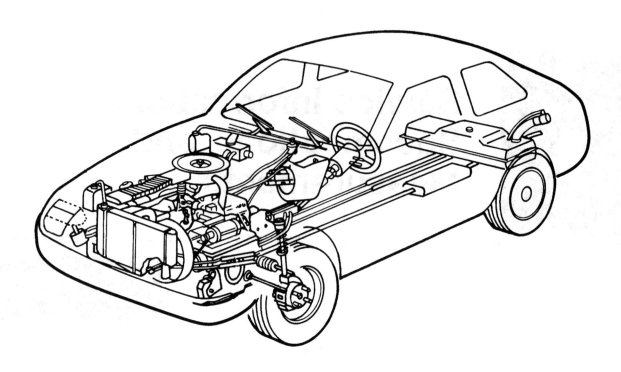

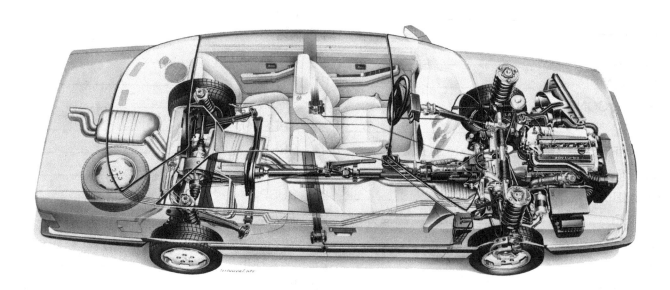

Name _____ Date _____ Instructor Review _____

Vehicle Identification Exercise

In the following figure, write down what each number or letter in the VIN designates.

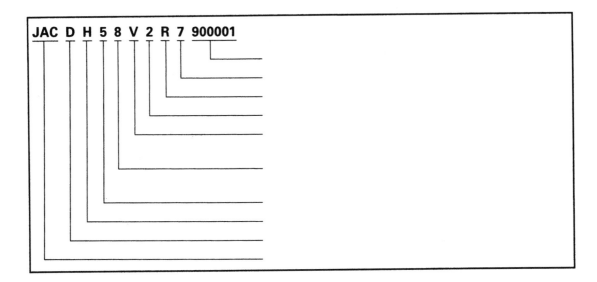

In the following figure, write in the VIN of a vehicle in your shop. In the spaces pointing to each number or letter, write down what information the numbers or letters represent.

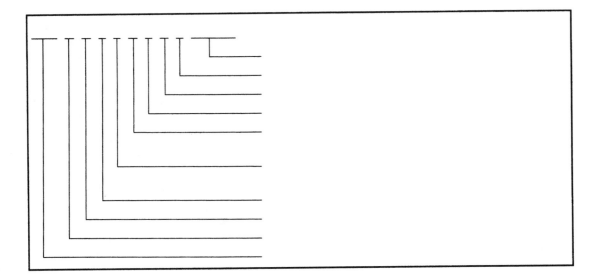

Name _____ Date _____ Instructor Review _____

Paint Mixing Instructions

Using the information listed next, answer the following questions. All questions refer to ChromaBase paint and clearcoat.

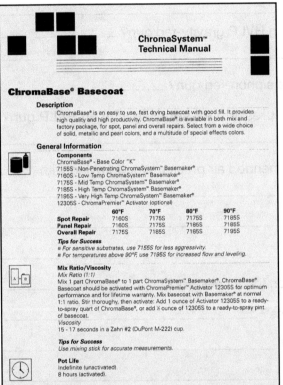

ChromaSystem™ Technical Manual

ChromaBase® Basecoat

Description

ChromaBase® is an easy to use, fast drying basecoat with good fill. It provides high quality and high productivity. ChromaBase® is available in both mix and factory package, for spot, panel and overall repairs. Select from a wide choice of solid, metallic and pearl colors, and a multitude of special effects colors.

General Information

Components
ChromaBase® - Base Color "K"
7155S - Non-Penetrating ChromaSystem™ Basemaker®
7160S - Low Temp ChromaSystem™ Basemaker®
7175S - Mid Temp ChromaSystem™ Basemaker®
7185S - High Temp ChromaSystem™ Basemaker®
7195S - Very High Temp ChromaSystem™ Basemaker®
12305S - ChromaPremier® Activator (optional)

	60°F	70°F	80°F	90°F
Spot Repair	7160S	7175S	7175S	7185S
Panel Repair	7160S	7175S	7185S	7185S
Overall Repair	7175S	7185S	7185S	7195S

Tips for Success
■ For sensitive substrates, use 7155S for less aggressivity.
■ For temperatures above 90°F, use 7195S for increased flow and leveling.

Mix Ratio/Viscosity
Mix Ratio (1:1)
Mix 1 part ChromaBase® to 1 part ChromaSystem™ Basemaker®. ChromaBase® Basecoat should be activated with ChromaPremier™ Activator 12305S for optimum performance and for lifetime warranty. Mix basecoat with Basemaker® at normal 1:1 ratio. Stir thoroughly, then activate: Add 1 ounce of Activator 12305S to a ready-to-spray quart of ChromaBase®, or add ½ ounce of 12305S to a ready-to-spray pint of basecoat.
Viscosity
15 - 17 seconds in a Zahn #2 (DuPont M-222) cup.

Tips for Success
Use mixing stick for accurate measurements.

Pot Life
Indefinite (unactivated).
8 hours (activated).

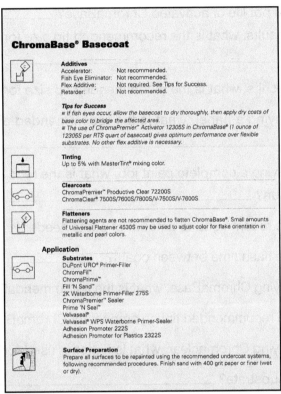

ChromaBase® Basecoat

Additives
Accelerator:	Not recommended.
Fish Eye Eliminator:	Not recommended.
Flex Additive:	Not required. See Tips for Success.
Retarder:	Not recommended.

Tips for Success
■ If fish eyes occur, allow the basecoat to dry thoroughly, then apply dry coats of base color to bridge the affected area.
■ The use of ChromaPremier™ Activator 12305S in ChromaBase® (1 ounce of 12305S per RTS quart of basecoat) gives optimum performance over flexible substrates. No other flex additive is necessary.

Tinting
Up to 5% with MasterTint® mixing color.

Clearcoats
ChromaPremier™ Productive Clear 72200S
ChromaClear® 7500S/7600S/7800S/V-7500S/V-7600S

Flatteners
Flattening agents are not recommended to flatten ChromaBase®. Small amounts of Universal Flattener 4530S may be used to adjust color for flake orientation in metallic and pearl colors.

Application

Substrates
DuPont URO® Primer-Filler
ChromaFil™
ChromaPrime™
Fill 'N Sand™
2K Waterborne Primer-Filler 275S
ChromaPremier™ Sealer
Prime 'N Seal™
Velvaseal®
Velvaseal® WPS Waterborne Primer-Sealer
Adhesion Promoter 222S
Adhesion Promoter for Plastics 2322S

Surface Preparation
Prepare all surfaces to be repainted using the recommended undercoat systems, following recommended procedures. Finish sand with 400 grit paper or finer (wet or dry).

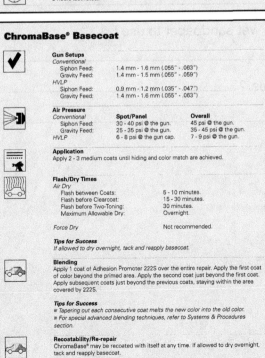

ChromaBase® Basecoat

Gun Setups
Conventional
Siphon Feed:	1.4 mm - 1.6 mm (.055" - .063")
Gravity Feed:	1.4 mm - 1.5 mm (.055" - .059")

HVLP
Siphon Feed:	0.9 mm - 1.2 mm (.035" - .047")
Gravity Feed:	1.4 mm - 1.6 mm (.055" - .063")

Air Pressure
Conventional	**Spot/Panel**	**Overall**
Siphon Feed:	30 - 40 psi @ the gun.	45 psi @ the gun.
Gravity Feed:	25 - 35 psi @ the gun.	35 - 45 psi @ the gun.
HVLP	6 - 8 psi @ the gun cap.	7 - 9 psi @ the gun.

Application
Apply 2 - 3 medium coats until hiding and color match are achieved.

Flash/Dry Times
Air Dry
Flash between Coats:	5 - 10 minutes.
Flash before Clearcoat:	15 - 30 minutes.
Flash before Two-Toning:	30 minutes.
Maximum Allowable Dry:	Overnight.

Force Dry   Not recommended.

Tips for Success
If allowed to dry overnight, tack and reapply basecoat.

Blending
Apply 1 coat of Adhesion Promoter 222S over the entire repair. Apply the first coat of color beyond the primed area. Apply the second coat just beyond the first coat. Apply subsequent coats just beyond the previous coats, staying within the area covered by 222S.

Tips for Success
■ Tapering out each consecutive coat melts the new color into the old color.
■ For special advanced blending techniques, refer to Systems & Procedures section.

Recoatability/Re-repair
ChromaBase® may be recoated with itself at any time. If allowed to dry overnight, tack and reapply basecoat.

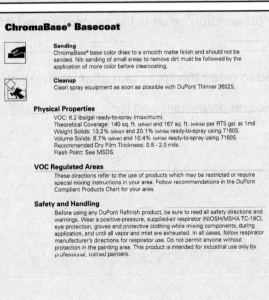

ChromaBase® Basecoat

Sanding
ChromaBase® base color dries to a smooth matte finish and should not be sanded. Nib sanding of small areas to remove dirt must be followed by the application of more color before clearcoating.

Cleanup
Clean spray equipment as soon as possible with DuPont Thinner 3602S.

Physical Properties
VOC: 6.2 lbs/gal ready-to-spray (maximum).
Theoretical Coverage: 140 sq. ft. (silver) and 167 sq. ft. (white) per RTS gal. at 1mil.
Weight Solids: 13.2% (silver) and 20.1% (white) ready-to-spray using 7160S.
Volume Solids: 8.7% (silver) and 10.4% (white) ready-to-spray using 7160S.
Recommended Dry Film Thickness: 0.5 - 2.0 mils.
Flash Point: See MSDS.

VOC Regulated Areas
These directions refer to the use of products which may be restricted or require special mixing instructions in your area. Follow recommendations in the DuPont Compliant Products Chart for your area.

Safety and Handling
Before using any DuPont Refinish product, be sure to read all safety directions and warnings. Wear a positive-pressure, supplied-air respirator (NIOSH/MSHA TC-19C), eye protection, gloves and protective clothing while mixing components, during application, and until all vapor and mist are exhausted. In all cases, follow respirator manufacturer's directions for respirator use. Do not permit anyone without protection in the painting area. This product is intended for industrial use only by professional, trained painters.

Courtesy of DuPont Automotive Finishes

1. Is ChromaBase a single- or two-stage paint? _____

2. Explain the difference between single- and two-stage paint. _____
   _____

3. When is 7165 reducer used? _____

4. When is 7185 reducer used? _____

5. What is the mix ratio? _____

6. What is the pot life of unactivated ChromaBase? _____

7. What is the pot life of activated ChromaBase? _____

8. For best results, what is the recommended tip size for an HVLP gravity-fed gun? _____
   _____

9. For best results, what is the recommended tip size for a siphon-fed gun? _____

10. When spraying a panel, what is the recommended air pressure for spraying with an HVLP gun?
    _____

11. When spraying a complete paint job, what is the recommended air pressure for spraying with
    an HVLP gun? _____

12. What is the recommended amount of coats needed? _____

13. What is the flash time between coats? _____

14. When spraying ChromaBase, what is the recommendation for accelerator? _____

15. What is the recommended tinting amount for ChromaBase? _____

16. When spraying Chromaclear, what is flattener used for? _____

17. What is a substrate? _____

18. For surface preparation, what is the recommended grit of wet sandpaper to use?
    _____

19. What is used to clean the spray gun after applying clearcoat? _____

20. What is adhesion promoter used for? _____

Name _____ Date _____

Metric and Standard Measuring

Objective
Upon completion of this activity sheet, the student should be able to make accurate measurements using a standard tape rule and a metric tape measure.

ASE Education Foundation Task Correlation
There are no ASE Education Foundation tasks related to or assessed in this job sheet.

We Support

ASE | **Education Foundation**

Tools and Equipment
Any vehicle
Standard and metric tape measure

Safety Equipment
Safety glasses or goggles

Introduction
Many modern vehicles use both standard and metric nuts and bolts. A good technician will have knowledge of both measuring systems.

Vehicle Description

Year _____ Make _____ Model _____

Task Completed ☐

Procedure
1. Accurate measurements are essential when checking for damage or when straightening a frame. Using the illustration provided for reference, take the measurements on your vehicle.

    a. Using your standard tape first, what is the distance between A and D? _____
    Metric? _____

    b. What part on the vehicle are you using for your measuring point? _____

    c. What is the distance between B and C? _____ Metric? _____

    d. Is the distance between A and D the same as between B and C? _____

    e. If not, what is the difference in inches? _____ In millimeters? _____

    f. If there is a difference, what does this indicate? _____

    g. If the measurements are not the same, close the hood and check the gaps between the left fender and the hood. What is the measurement in inches? _____ In millimeters? _____

    h. Repeat the same for the right side of the vehicle. In inches? _____ In millimeters? _____

    i. What is the difference between C and F in inches? _____ In millimeters? _____

    j. What measuring point are you using when measuring around the windshield? _____

    k. What is the difference between D and E in inches? _____ In millimeters? _____

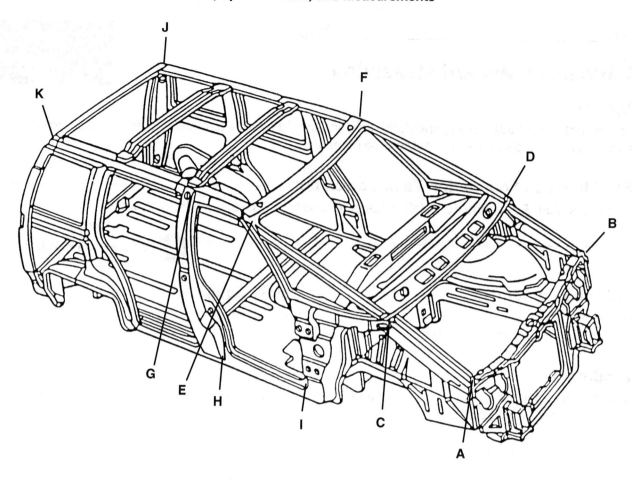

l.   Is the distance between C and F the same as between D and E? _____

m. If not, what is the difference in inches? _____ In millimeters? _____

n.   If there is a difference, what does this indicate? _____

o.   What is the distance between E and H in inches? _____ In millimeters? _____

p.   What is the distance between I and G in inches? _____ In millimeters? _____

q.   Are these distances the same as for the same panel on the opposite side of the vehicle?
_____

r.   If not, what does this indicate? _____

s.   If not, what is the difference in inches? _____ In millimeters? _____

t.   What is the distance between E and J in inches? _____ In millimeters? _____

u.   What is the distance between F and K in inches? _____ In millimeters? _____

v.   Is the distance between E and J the same as that between F and K? _____

w. If not, what is the difference in inches? _____ In millimeters? _____

x.   If not, what does this indicate? _____

INSTRUCTOR'S COMMENTS _____
_____
_____

Review Questions

Name _____ Date _____ Instructor Review _____

1. VIN is an abbreviation for _____.

2. _____ is an abbreviation for cruise control.

3. A(n) _____ should always be worn when spraying ChromaBase.

4. Using CDs is still the best way to get the most current collision repair procedure information.
   A. True
   B. False

5. The year of a vehicle is encoded in the VIN.
   A. True
   B. False

6. Paint reduced 250 percent is almost transparent.
   A. True
   B. False

7. Technician A says that a 2:1 ratio is a reduction of 50 percent. Technician B says that a ratio of 2:1 is a reduction of 100 percent. Who is correct?
   A. Technician A
   B. Technician B
   C. Both Technician A and Technician B
   D. Neither Technician A nor Technician B

8. Collision repair procedures give instructions, safety warnings, and technical illustrations for specific makes and models of vehicles.
   A. True
   B. False

9. Technician A says that because ChromaBase does not take a catalyst, a dust mask is sufficient protection when spraying. Technician B says that an approved respirator must be worn when spraying ChromaBase. Who is correct?
   A. Technician A
   B. Technician B
   C. Both Technician A and Technician B
   D. Neither Technician A nor Technician B

10. Technician A says that the paint code can be found in the VIN. Technician B says that it cannot. Who is correct?
   A. Technician A
   B. Technician B
   C. Both Technician A and Technician B
   D. Neither Technician A nor Technician B

Hand Tools

Name _____ Date _____ Instructor Review _____

Hand Tool Identification

1. Hammers—For each type of hammer, give the name and explain the situation in which it is used.

a. Name _____
   Use _____
   _____

b. Name _____
   Use _____
   _____

c. Name _____
   Use _____
   _____

d. Name _____
   Use _____
   _____

e. Name _____
   Use _____
   _____

f. Name _____
   Use _____
   _____

2. Dollies—For each type of dolly, give the name and identify which type of panel it can be used to repair.

a. Name _____

   Panel _____

b. Name _____

   Panel _____

c. Name _____

   Panel _____

3. Spoons—For each spoon, give the name and its use.

a. Name _____

   Use _____

   _____

b. Name _____

   Use _____

   _____

c. Name _____

   Use _____

   _____

4. Pullers—For each type of puller, give the name and its use.

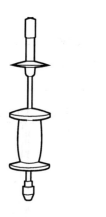   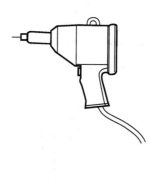   

a. Name _____

   Use _____

   _____

b. Name _____

   Use _____

   _____

c. Name _____

   Use _____

   _____

5.   Sanders—For each type of sander, give the name, its length, and its use.

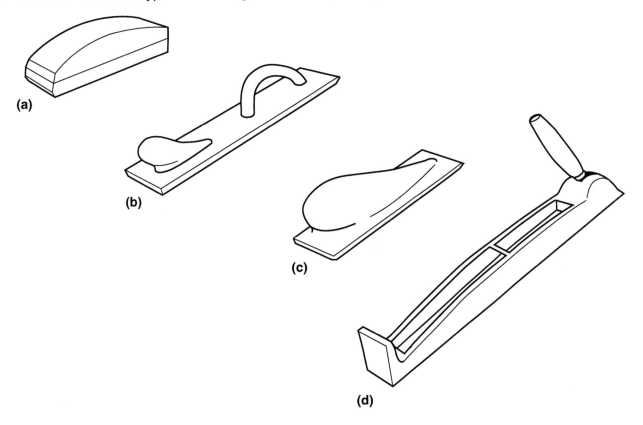

(a)

(b)

(c)

(d)

a. Name _____

   Length _____

   Use _____

   _____

b. Name _____

   Length _____

   Use _____

   _____

c. Name _____

   Length _____

   Use _____

   _____

d. Name _____

   Length _____

   Use _____

   _____

Name _____ Date _____ Instructor Review _____

Uses of Hand Tools

Write the name and use of each tool shown in the illustrations next to the appropriate letters.

A. _____

  Use _____

  _____

B. _____

  Use _____

  _____

C. _____

  Use _____

  _____

D. _____

  Use _____

  _____

E. _____

  Use _____

  _____

F. _____

  Use _____

  _____

G. _____

Use _____

_____

H. _____

Use _____

_____

I. _____

Use _____

_____

A
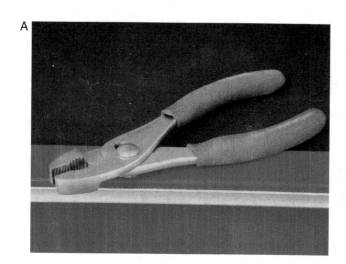

B
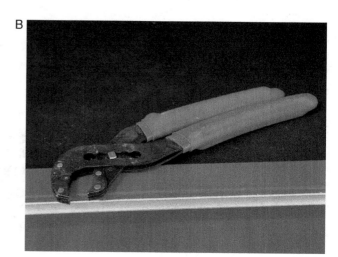

C
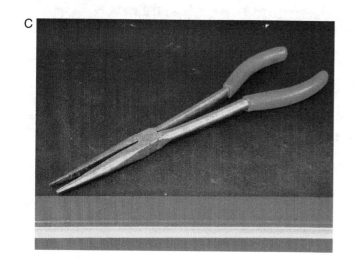

D
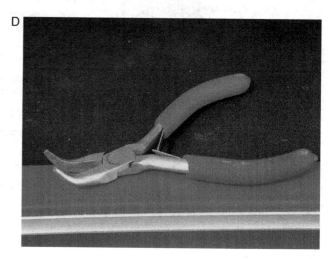

E

D. _____

　Use _____

　_____

E. _____

　Use _____

　_____

A. _____

　Use _____

　_____

B. _____

　Use _____

　_____

C. _____

　Use _____

　_____

Name _____ Date _____

Hammer and Dolly Use

Objective

Upon completion of this activity sheet, the student should be able to safely and accurately use a hammer and dolly.

ASE Education Foundation Task Correlation

II.D.2   Locate and repair surface irregularities on a damaged body panel using power tools, hand tools, and weld-on pulling attachments. **(HP-I)**

II.D.3   Demonstrate hammer and dolly techniques. **(HP-I)**

We Support

 | **Education Foundation**

Tools and Equipment

Pick hammer
Fender
Dollies
Ruler or tape measure

Safety Equipment

Safety glasses or goggles

Introduction

The hammer and dolly are probably the most important tools that a technician uses. When doing metal work, the proper use of the hammer and dolly determines the quality of the finished work. Even with all the changes in paints, plastics, and other materials, proper hammer and dolly work has remained a constant. Accuracy is critical because a technician works around windshields, moldings, and other expensive parts that must not be damaged.

Procedure

Hammer

Task Completed

1. Go to one of the two metal workbenches in the shop, and put masking tape down on a 2-foot × 2-foot area on the bench. Draw 10½-inch diameter circles.   ☐

2. Use a pick hammer, and try to hit the center of the circles. Start with hitting each circle 10 times. Remember, you should be using a light bouncing motion. **DO NOT** use the hammer as if you were driving nails.   ☐

   a.   Out of all 10 circles, how many strikes were out of the circle?

   _____

   b.   Were some blows deeper than others?

   _____

3. Draw another ½-inch circle or two, and continue practicing until you can consistently hit nine out of ten tries inside the circle.   ☐

4. Make 10 more circles on the bench; this time ¼ inch in diameter.   ☐

Task Completed

5. Use the pick hammer to hit inside the circle as before. Practice until you can consistently hit nine out of ten tries inside the circle. ☐

    a. Out of all 10 circles, how many strikes were out of the circle?

_____

    b. Were some blows deeper than others?

_____

6. Draw 10 Xs 2 inches tall on the bench. ☐

7. Practice hitting the center of each X 10 times. ☐

    a. How many times did you hit the X?

_____

    b. How many times did you miss?

_____

8. Practice hitting the tip of each X 10 times. ☐

    a. How many times did you hit the tip?

_____

    b. How many times did you miss?

_____

Dolly

1. Draw 10 circles that are 1 inch in diameter this time on a test panel assigned to you by the instructor. Practice hitting the circle using the *hammer-on-dolly* method. ☐

    a. What is an easy way to tell whether you are hitting directly on the dolly?

_____

_____

    b. Explain when the hammer-on-dolly process is used or why.

_____

_____

2. Practice hitting the circle using the *hammer-off-dolly* method. ☐

    a. What is the purpose of the hammer-off-dolly method?

_____

_____

INSTRUCTOR'S COMMENTS _____

_____

_____

Review Questions

Name _____ Date _____ Instructor Review _____

1. Technician A buys tools with a lifetime warranty. Technician B buys the cheapest tools he can find. Who is correct?
   A. Technician A
   B. Technician B
   C. Both Technician A and Technician B
   D. Neither Technician A nor Technician B

2. Technician A uses an adjustable wrench on all nuts. Technician B uses a combination wrench when possible. Who is correct?
   A. Technician A
   B. Technician B
   C. Both Technician A and Technician B
   D. Neither Technician A nor Technician B

3. A _____-inch long pipe wrench is adequate for most body shop applications.

4. A Torx fastener is also called a _____ _____.

5. Technician A states that a socket and ratchet is faster than a wrench. Technician B believes that a wrench is faster than a socket and ratchet. Who is correct?
   A. Technician A
   B. Technician B
   C. Both Technician A and Technician B
   D. Neither Technician A nor Technician B

6. Technician A states that a slotted head screw slips less during removal than a Phillips head screw. Technician B believes that a Phillips head screw slips less. Who is correct?
   A. Technician A
   B. Technician B
   C. Both Technician A and Technician B
   D. Neither Technician A nor Technician B

7. Technician A uses a small slotted screwdriver to turn a Phillips head screw. Technician B uses a Phillips screwdriver only to turn a Phillips head screw. Who is correct?
   A. Technician A
   B. Technician B
   C. Both Technician A and Technician B
   D. Neither Technician A nor Technician B

8. Adjustable pliers are also called _____ _____.

9. Vise grips can be used for clamping metal together and for turning rounded off fasteners.
   A. True
   B. False

10. Technician A uses a cheater pipe to help turn off a rusted nut. Technician B heats the nut, and then turns it off with a ratchet. Who is correct?
   A. Technician A
   B. Technician B
   C. Both Technician A and Technician B
   D. Neither Technician A nor Technician B

Power Tools and Equipment

Name _____ Date _____ Instructor Review _____

Power Tool Identification Lab

1. Power Tools—For each type of power tool, give the name and explain the situation in which it is used.

a. Name _____

   Use _____

   _____

b. Name _____

   Use _____

   _____

c. Name _____

   Use _____

   _____

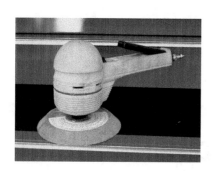

d. Name _____

   Use _____

   _____

e. Name _____

   Use _____

   _____

f. Name _____

   Use _____

   _____

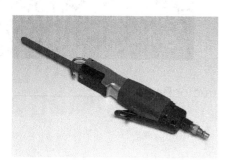

g. Name _____

   Use _____

   _____

h. Name _____

   Use _____

   _____

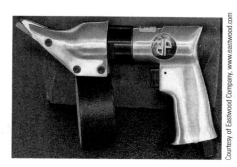

i. Name _____

   Use _____

   _____

j. Name _____

   Use _____

   _____

k. Name _____

   Use _____

   _____

l. Name _____

   Use _____

_____

m. Name _____

    Use _____

_____

Courtesy of SPX/OTC Service Solutions

n. Name _____

   Use _____

_____

o. Name _____

   Use _____

_____

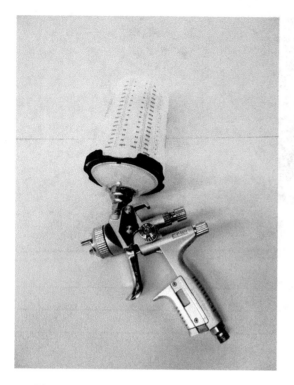

p. Name _____

   Use _____

   _____

Chief Automotive Technologies

r. Name _____

   Use _____

   _____

Courtesy of SPX/OTC Service Solutions

Courtesy of SPX/OTC Service Solutions

q. Name _____

   Use _____

   _____

s. Name _____

   Use _____

   _____

t. Name _____

   Use _____

   _____

Name _____ Date _____

Power Tool Operation

Objective

Upon completion of this activity sheet, the student should be able to identify and safely use auto body power tools.

ASE Education Foundation Task Correlation

II.D.1   Prepare a panel for body filler by abrading or removing the coatings; featheredge and refine scratches before the application of body filler. **(HP-I)**

We Support

ASE | Education Foundation

Tools and Equipment

Grinder
Air file
8-inch sander
DA sander
Painted fender

Safety Equipment

Safety glasses or goggles
Dust respirator
Work gloves

Introduction

Power tools are extremely useful in helping the technician produce quality work in an efficient and a timely manner. These tools can save many labor hours because of their speed but can also create damage very quickly if they are not used properly and carefully. Proper maintenance is also very important because power tools are expensive to replace.

Procedure

Task Completed

1. Check the condition of the pad, disc, or paper before operation. If any paint buildup, nicks, tears, or holes are found, make the necessary repairs. _____   ☐

DA Sander

2. Obtain a DA sander.   ☐

    a.   What does DA stand for? _____

    b.   Explain the importance of a clean pad when applying sandpaper to a DA.

    _____

    _____

    c.   Explain the two purposes of a DA.

    _____

    d.   What are the two most commonly used sandpaper grits, and what are their uses?

    _____

    e.   How often should this tool be oiled?

    _____

3. With a nail or a screwdriver, make a scratch of about 10 inches in your panel. Using
#180 grit, place the sander flat, then pitch it at a very slight angle, and start sanding
in an up-and-down motion. Make sure the speed is not set too high. Keep sanding
until the paint is featheredged out about 2 or 3 inches. At this point you will see all
the layers of paint, primer, and whatever sealers may be on the panel.

☐

    a.   Does your repair area featheredge out smooth, or does it still feel wavy?

    _____

    b.   How will you correct this if it is wavy?

    _____

    c.   If it is straight, what grit sandpaper should you finish sand the area with?

    _____

Grinder

4. Draw another 5-inch × 5-inch square on your panel. Your grinder should have a
rough grit such as #24 or #36 grit disc on it. Hold the grinder at a slight angle and
grind side to side; then grind up and down.

☐

    a.   What safety precautions must be taken when using the grinder?

    _____

5. When grinding, make sure you do not stay in one spot too long or the metal can
be warped by extreme heat. As an experiment, grind on one spot without moving.
Notice how quickly the metal turns blue and then red and then how quickly a hole
appears in the metal. This must not happen when you work on a vehicle because the
cost to replace the part could be deducted from your paycheck if your boss sees it
as negligence.

☐

    a.   How long did it take to burn a hole in the metal? _____

    b.   Explain why it is important to stop the grinder on the panel when stopping the trigger.

    _____

    _____

    _____

8-Inch Sander

6. Also known as a "mud buster" or "mud hog," the 8-inch sander is used to quickly
remove material from a panel. It is important to keep in mind that this sander is not a
finishing tool. However, it is ideal for stripping large areas of paint.

    a.   What grit is commonly used for this tool? _____

    b.   Draw a 12-inch × 12-inch square on a flat panel. Hold the sander flat while
sanding. Use the sander in a side-to-side motion; then repeat using an
up-and-down motion. Notice how quickly the paint or plastic filler is removed.

    c.   What is the difference in bare metal when using an 8-inch sander compared to a
grinder? _____

    _____

    _____

INSTRUCTOR'S COMMENTS _____

_____

_____

Review Questions

Name _____ Date _____ Instructor Review _____

1. Technician A says that battery-powered tools are becoming more common than electric and air tools because they do not require having any cords or hoses to drag across the floor. Technician B says it is much better to use electric tools as the battery charge times are too long. Who is correct?
   A. Technician A
   B. Technician B
   C. Both Technician A and Technician B
   D. Neither Technician A nor Technician B

2. Technician A uses only impact sockets on his impact wrench. Technician B believes that for small bolts a chrome socket is an acceptble choice to use on an impact wrench. Who is correct?
   A. Technician A
   B. Technician B
   C. Both Technician A and Technician B
   D. Neither Technician A nor Technician B

3. Technician A uses an impact wrench to final torque a fastener. Technician B uses a torque wrench. Who is correct?
   A. Technician A
   B. Technician B
   C. Both Technician A and Technician B
   D. Neither Technician A nor Technician B

4. The "walking" of a spinning drill can be prevented by first using a _____.

5. Technician A says that white foam pads are typically used for compounding. Technician B says that black foam pads are typically used for polishing and swirl removing. Who is correct?
   A. Technician A
   B. Technician B
   C. Both Technician A and Techniclan B
   D. Neither Technician A nor Technician B

6. The two basic types of air sanders are _____ and _____.

7. Technician A says that hook and loop sandpaper is more reusable than stick it sandpaper. Technician B says that stick it sandpaper is more reusable than hookit sandpaper. Who is correct?
   A. Technician A
   B. Technician B
   C. Both Technician A and Technician B
   D. Neither Technician A nor Technician B

8. Technician A uses a heat gun to reshape plastic parts. Technician B uses a heat gun to dry paint. Who is correct?
   A. Technician A
   B. Technician B
   C. Both Technician A and Technician B
   D. Neither Technician A nor Technician B

9. A belt sander is mainly used to grind down spot welds for faster welded panel removal.
   A. True
   B. False

10. Which lift is best for estimating body damage?
    A. Four-post and two-post
    B. Side-post
    C. Center-post

CHAPTER 7

Compressed Air System Technology

Name _____ Date _____

Compressed Air Systems

Objective
Upon completion of this activity sheet, the student should be able to identify and safely use compressed air systems.

ASE Education Foundation Task Correlation
This activity sheet will help the student become more proficient at ASE Education Foundation learning skills pertaining to maintaining a safe and healthy environment.

We Support
ASE | **Education Foundation**

Tools and Equipment
Any compressed air system
Air pressure gauge
Tape measure

Safety Equipment
Safety glasses or goggles

Introduction
In the auto body shop, the compressor is the "pumping heart" that keeps all pneumatic tools moving. Without it, the shop will come to a complete halt. When properly maintained, the compressor will provide the shop with trouble-free service for years. Preventive maintenance is performed by shop employees. Any necessary major repairs are done by an outside specialist.

1. What is the brand of compressor in your shop? _____

2. What is the age of the compressor? _____

3. With the help of your instructor, determine the cost of replacing this system. _____

Procedure

4. Make an outline drawing of your shop's air system in the space provided. In the diagram, include the air compressor, air lines (including the total number of lines and their length and inner diameters), air transformers, filters, driers, water traps, and air hoses.   ☐

5. What is the line pressure at each air transformer?   ☐

           Transformer   Line Pressure

      _____   _____

      _____   _____

      _____   _____

6. Attach a 25-foot air hose to an air transformer.   ☐
   a. What is the air pressure at the end of the hose? _____

7. Attach two more 25-foot hoses with tools in operation.   ☐
   a. With the tools running, is the air pressure in the first hose the same as in step 6a?
   b. If not, how much has it dropped?_____
   c. What does this indicate? _____
   d. If there is a pressure drop, explain how this could affect the quality of your work.

       _____
   e. What is the air pressure at the transformer?_____
   f. Why is there a difference? _____

8. With your instructor present, inspect the compressor in your shop. **DO NOT TOUCH** anything on this compressor whether or not it is in operation. Hands, fingers, loose clothing, or dangling jewelry can get caught in the moving parts.
   a. Make a drawing of your school's air compressor in the space provided. Include the air intake, cylinders, electric motor, tank (indicate size), pressure cutoff switch, pressure gauge, tank drain, and drive belt.

☐

9. Your instructor will explain and demonstrate how to replace a damaged hose end. Take notes while this is done. List each step in the process in the space provided, paying very close attention to safety because compressed air can be dangerous.

☐

   a. _____
   b. _____
   c. _____
   d. _____

   e. _____
   f. _____
   g. _____
   h. _____

INSTRUCTOR'S COMMENTS _____
_____
_____

Review Questions

Name _____ Date _____ Instructor Review _____

1. The three types of air compressors are _____, _____, and _____.

2. Which compressor is most efficient?
   A. Single-stage
   B. Two-stage

3. Which is the best rating to use when selecting an air compressor?
   A. Displacement cfm
   B. Free air cfm

4. Air compressor tanks can be mounted vertically or _____.

5. Technician A drains the air compressor tank each day. Technician B drains the air compressor tank each month. Who is correct?
   A. Technician A
   B. Technician B
   C. Both Technician A and Technician B
   D. Neither Technician A nor Technician B

6. The _____ starts and stops the compressor motor based on the system pressure.

7. An air transformer is also called a _____.

8. Elimination of oil, dirt, and moisture in air lines is provided by a _____.

9. Technician A uses an aftercooler to remove moisture. Technician B believes that the aftercooler removes oil. Who is correct?
   A. Technician A
   B. Technician B
   C. Both Technician A and Technician B
   D. Neither Technician A nor Technician B

10. A ¼-inch hose is acceptable for all uses in a body shop.
   A. True
   B. False

Collision Repair Materials and Fasteners

Name _____ Date _____ Instructor Review _____

Common Screw Identification

Give the proper name and uses for each screw shown in the illustration.

A. _____ Uses _____
B. _____ Uses _____
C. _____ Uses _____
D. _____ Uses _____
E. _____ Uses _____
F. _____ Uses _____
G. _____ Uses _____
H. _____ Uses _____
I. _____ Uses _____
J. _____ Uses _____
K. _____ Uses _____
L. _____ Uses _____
M. _____ Uses _____
N. _____ Uses _____
O. _____ Uses _____

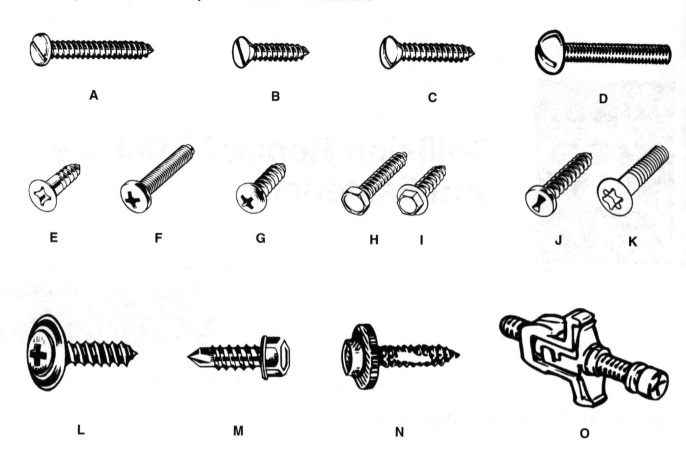

A

B

C

D

E

F

G

H

I

J

K

L

M

N

O

Name _____ Date _____ Instructor Review _____

Common Abrasive Uses

Identify the sandpapers shown in the following illustration and give their uses.

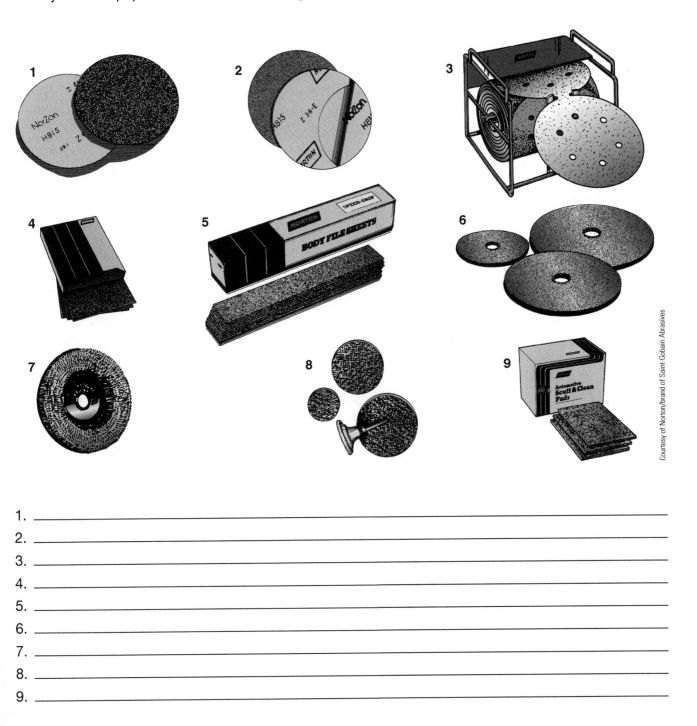

Courtesy of Norton/brand of Saint-Gobain Abrasives

1. _____
2. _____
3. _____
4. _____
5. _____
6. _____
7. _____
8. _____
9. _____

Name _____ Date _____

Fastener Identification and Removal

Objective

Upon completion of this activity sheet, the student should be able to identify and safely remove some of the more common fasteners in automobiles.

ASE Education Foundation Task Correlation

There are no ASE Education Foundation tasks related to or assessed in this job sheet.

We Support

Education Foundation

Tools and Equipment

Various trim removal tools
Vehicle
Various sockets and ratchets
Assorted bit set
Various screwdrivers

Safety Equipment

Safety glasses or goggles
Dust respirator
Work gloves

Introduction

There are so many different types of fasteners used in today's automobiles. In this exercise, you will learn to use different tools to remove and replace some of these types of fasteners. Keep in mind that these instructions are very generic, and they will differ from vehicle to vehicle.

Vehicle Description

Year _____ Make _____ Model _____

VIN _____

	Task Completed

Procedure

1. Have your instructor assign you to an automobile to remove some different fasteners. Most newer automobiles have the most amount of different fasteners in the front or rear bumpers. ☐

2. Look at the front bumper cover. Note the different fasteners used to fasten the bumper cover to the vehicle. You should see push-in clip retainers underneath the cover (sometimes there are both push-in clips and screws or bolts going into a threaded clip underneath also) and sometimes across the top of the cover when you open the hood. Remove these with the most appropriate clip removal tool. ☐

   a. What types of fasteners did you remove? _____
   _____

   b. What tool or tools did you use? _____
   _____

c. How many push-in clips, screws, or bolts did you remove?

_____

_____

d. How many broke?

_____

3. Look inside the inner fender area where the bumper cover meets the fender. You ☐
should see a 10-mm bolt head, or sometimes it can be a coarse threaded Philips
head screw. Remove these with the appropriate tool.

a. What tool did you use? _____

4. Show your instructor what you removed so they can check your work and make ☐
sure nothing was damaged.

5. Once they have seen these fasteners, replace them in the reverse order that you ☐
removed them.

Now, repeat the same steps on the rear bumper cover.

6. Note the different fasteners used to fasten the bumper cover to the vehicle. You ☐
should see push-in clip retainers underneath the cover (sometimes there are both
push-in clips and screws or bolts going into a threaded clip underneath also) and
sometimes across the top of the cover when you open the trunk. Remove these
with the most appropriate clip removal tool.

a. What all types of fasteners did you remove? _____

_____

b. What tool or tools did you use? _____

_____

c. How many push-in clips, screws, or bolts did you remove?

_____

d. How many broke? _____

7. Look inside the inner fender area where the bumper cover meets the quarter panel. ☐
You should see a 10-mm bolt head, or sometimes it can be a coarse threaded Philips head screw. Remove these with the most appropriate tool. Sometimes you may
have a mud flap on the lower part of the bumper cover. If so, remove the necessary
fasteners from them as well.

a. What all types of fasteners did you remove? _____

_____

b. What tool or tools did you use to remove these?

_____

8. Show your instructor what all you removed so they can check your work and make ☐
sure nothing was damaged.

9. Once they have seen these fasteners, replace them in the reverse order that you ☐
removed them.

INSTRUCTOR'S COMMENTS _____

_____

_____

Review Questions

Name _____ Date _____ Instructor Review _____

1. #400 grit is a _____ grit sandpaper.

2. Smooth transition tape is made for undetectable tape lines on hard edge surfaces.
   A. True
   B. False

3. A flex agent is used when spraying a _____.

4. Epoxy primers can be sprayed over bare metal.
   A. True
   B. False

5. Two-sided tape is used for holding decals and pinstripe to automobiles.
   A. True
   B. False

6. Antichip coating is also called chip guard.
   A. True
   B. False

7. Technician A says that the higher the number, the finer the grit. Technician B says that the higher the number, the rougher the grit. Who is correct?
   A. Technician A
   B. Technician B
   C. Both Technician A and Technician B
   D. Neither Technician A nor Technician B

8. Technician A says that a self-tapping screw needs a nut to hold it in place. Technician B says that a nut is not needed. Who is correct?
   A. Technician A
   B. Technician B
   C. Both Technician A and Technician B
   D. Neither Technician A nor Technician B

9. Technician A says that #400 grit paper must be used wet. Technician B says that #400 grit can be used dry. Who is correct?
   A. Technician A
   B. Technician B
   C. Both Technician A and Technician B
   D. Neither Technician A nor Technician B

10. Technician A says that fish-eye eliminator should be used all the time as a precaution. Technician B says that fish-eye eliminator should only be used when necessary. Who is correct?
    A. Technician A
    B. Technician B
    C. Both Technician A and Technician B
    D. Neither Technician A nor Technician B

Outer Body Panel Service

Name _____ Date _____ Instructor Review _____

Bumper Assembly Identification

Using the following figure, name the parts and their functions.

1. _____

2. _____

3. _____

4. _____

5. _____

6. _____

7. _____

8. _____

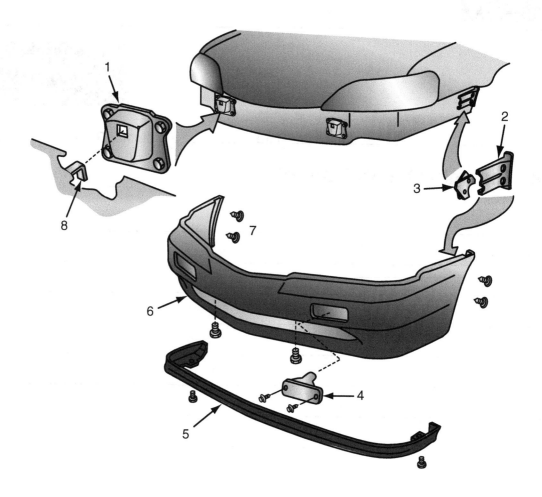

Name _____ Date _____

Remove and Install a Hood

Objective

Upon completion of this activity sheet, the student should be able to safely inspect, remove, install, and align any hood.

ASE Education Foundation Task Correlation

II.A.3   Identify vehicle system hazard types (supplemental restraint system [SRS], hybrid/electric/ alternative fuel vehicles), locations, and recommended procedures before inspecting or replacing components. **(HP-I)**

II.B.1   Review damage report and analyze damage to determine appropriate methods for overall repair; develop and document a repair plan. **(HP-I)**

II.B.4   Inspect, remove, label, store, and reinstall body panels and components that may interfere with or be damaged during repair. **(HP-I)**

II.C.1   Inspect/locate direct, indirect, or hidden damage and direction of impact. **(HP-I)**

II.C.2   Inspect, remove, and replace mechanically fastened welded steel panel or panel assemblies. **(HP-G)**

II.C.3   Determine the extent of damage to aluminum body panels; repair or replace. **(HP-G)**

II.C.4   Inspect, remove, replace, and align hood, hood hinges, and hood latch. **(HP-I)**

We Support

ASE | **Education Foundation**

Tools and Equipment

Any vehicle
Appropriate ratchets, sockets, and specialty tools
Tape measure
Tape

Safety Equipment

Safety glasses or goggles

Introduction

Hoods are the largest and heaviest panels on a vehicle. Hood panels cover the engine compartments on most vehicles. On rear engine vehicles, the hood serves as a trunk lid. With ongoing maintenance of the engine, the hood is opened and closed thousands of times during the life of a vehicle. If you have any difficulty with this learning exercise, be sure to ask your instructor for help. When working with hoods, always have two partners help you because the windshield can easily be broken if the hood slips.

Vehicle Description

Year _____ Make _____ Model _____

VIN _____

Procedure

Removing the Hood

1. Before starting work on the hood, wash and dry it to see whether any damage is present.   ☐

2. Measure the gaps between all adjoining panels and the hood.   ☐

    a. Write down your measurements; a good practice is to take a picture of the gaps for reference later. _____

3. Check the operation of the hinges and lock mechanism. If the hinges are not working properly, excessive pressure will be applied to the hood, causing damage.   ☐

    a. Does the hood close easily? _____

    b. If not, do the hinges and lock mechanism need lubrication? _____

4. Place a towel or fender cover under the hood to prevent the hood from damaging the paint, cowl panel, or windshield. Apply two or three strips of a 2-inch tape along the top edges of the fenders and wiper cowl panel.   ☐

    a. What does the tape help protect the vehicle from? _____

5. With the hood open, carefully disconnect the wire for the hood light (if present). Next, disconnect the windshield washer hoses if they are run through the hood. Always remember to twist the hoses to free them up before you just pull on them.   ☐

    a. Is there an insulation pad covering the underside of the hood? _____

    b. What is the purpose of this pad? _____

    _____

    c. This pad is held in with large plastic retaining clips. Make sure to label the clips in a Ziploc bag to store them so they do not get lost or mixed up. What specialty tool is used to remove these clips? _____

    d. Did you break any? _____

    If so, please write down the number of broken clips and tell the instructor how many. _____

6. Remove the insulator pad and place on a parts rack. Be careful not to rip or tear it.   ☐

    a. What is the condition of the pad? _____

7. Carefully mark the underside of the hood with tape around the hinges, and then trace an outline of the hinge on the tape. **DO NOT** draw on the hood itself.   ☐

    a. What is the reason for marking the hinges? _____

    _____

8. Remove the hood latch assembly; this will make initial adjustments easier when reinstalling the hood. Sometimes you will need to remove the grille to access the latch.   ☐

9. With the help of two partners, remove the front bolt on each hinge holding the   ☐
hood. Position one person at each hinge, while one person holds the front of the
hood up. It is much easier if each person at the hinge has his or her own ratchet.

   a. Why are the front bolts removed before the back bolts? _____

   _____

10. Before loosening the rear bolts, make sure the person at the hinges holds the   ☐
bottom of the hood with one hand while loosening the bolt with the other free
hand. It is also a good idea to hold the hood up with your shoulder while you
remove the rear bolts. If there is too much pressure on the bolts, damage can
occur. With all the bolts out, carefully lift the hood up and away from the car.
Place the hood upside down on a *clean*, padded workbench.

   a. Why is it so important to have a clean, padded workbench? _____

   _____

11. Use a good all-purpose cleaner to clean the inside of the hood, removing all   ☐
grease and dirt. Make sure none of the edges is dirty, because paint will not
adhere to oil, grease, wax silicone, and so on.

   a. Are there any factory stickers or tags on the panel? _____

   b. If so, what information is listed on these stickers? _____

Reinstalling the Hood

12. With the help of two partners, place the hood over the engine compartment.   ☐
Each partner should hold the hood with one hand in the middle of one side. The
other hand should hold the bottom corner where the hood meets the windshield.
Have your partners lift the hood up while you insert the back bolts in the hood
hinge. Align the hinges with your tape markings on the hood. **TRY** to align the
bolt washers with the original markings.

   a. Why is it important for your partners to hold the corner of the hood? _____

   _____

   b. Why is it important to insert the back bolts first? _____

   _____

13. With the back bolts hand-tightened, slowly close the hood and make sure to pay   ☐
attention to the fender edges and check all gaps. You **DO NOT** want to scratch
or chip any paint on the surrounding panels. It is not very often that the hood will
line up on the first try so do not get frustrated. Keep making necessary adjustments to obtain straight and even gaps all around.

14. Once the hood is properly aligned, measure all the gaps again.   ☐

   a. Are all the gaps the same as they were before the hood was removed?

   _____. If not, make the necessary adjustments to get them back to
   original.

Task Completed

15. After all the gaps are equal, make sure the latch and hinges are greased   ☐
    (if needed). Wash the hood and/or pre-clean with wax and grease remover to
    remove any grease or handprints. Check for damage that might have occurred
    during this operation, and report it to your instructor.

    a. List any damage if applicable: _____

    _____

16. Once the latch is adjusted properly, adjust one of the end rubber grommets   ☐
    (either on the underside of the hood or on the radiator support) all the way out if it
    has them. If it does not, disregard 15a. and 16a.

    a. How does this affect the height of the deck lid? _____

    _____

17. Adjust one of the grommets all the way in.   ☐

    a. How does this affect the height of the deck lid? _____

    _____

INSTRUCTOR'S COMMENTS _____

_____

_____

Name _____ Date _____

Remove and Install a Deck Lid/Trunk

Objective

Upon completion of this activity sheet, the student should be able to inspect, remove, replace, and align any deck lid within the time set by the instructor.

ASE Education Foundation Task Correlation

II.A.1	Select and use proper personal safety equipment; take necessary precautions with hazardous operations and materials in accordance with federal, state, and local regulations. **(HP-I)**
II.A.2	Locate procedures and precautions that may apply to the vehicle being repaired. **(HP-I)**
II.A.3	Identify vehicle system hazard types (supplemental restraint system [SRS], hybrid/electric/alternative fuel vehicles), locations, and recommended procedures before inspecting or replacing components. **(HP-I)**
II.C.1	Inspect/locate direct, indirect, or hidden damage and direction of impact. **(HP-I)**
II.C.2	Inspect, remove, and replace mechanically fastened welded steel panel or panel assemblies. **(HP-G)**
II.C.3	Determine the extent of damage to aluminum body panels; repair or replace. **(HP-G)**
II.C.5	Inspect, remove, replace, and align deck lid, lid hinges, and lid latch. **(HP-I)**

We Support

ASE | Education Foundation

Tools and Equipment

Vehicle
Appropriate screwdrivers, ratchets, and sockets
Masking tape
Flashlight
Tape measure
Water hose

Safety Equipment

Safety glasses or goggles
Work gloves

Introduction

Deck lids are often repaired, removed, and replaced during body shop operations. These panels must open and close thousands of times during the life span of a vehicle. Horizontal panels are more susceptible to water leaks, so alignment is critical. If you have any difficulty or doubt about any aspect of this learning activity, be sure to ask your instructor for help or to answer your questions.

Vehicle Description

Year _____ Make _____ Model _____

VIN _____

Procedure

1. Before starting work, check the operation of the deck lid. ☐

   a. Does it close and latch with minimal effort?

   _____

2. With deck lid closed and latched, measure the gaps or a good practice to always ☐
   do is to take a picture for reference later when reinstalling.

   a. Write down your measurements of each gap width. _____

   _____

   b. To prevent the deck lid/trunk from damaging the paint when being removed,
   apply two or three strips of a 2-inch tape along the top edges of the quarter
   panels, taillights, and bumper cover if necessary.

   c. What is the reason for this? _____

3. Disconnect any wires attached to or running through the deck lid braces. ☐

   a. What is the function of these wires in the lid? _____

   _____

4. Carefully mark the underside of the deck lid with tape around the hinges and then ☐
   trace an outline of the hinge on the tape. **DO NOT** draw on the deck lid itself.

   a. What is the purpose of this? _____

   _____

5. To remove the latch in the deck lid, most of the time it has two or three bolts in it and ☐
   a latch cable. Carefully remove the cable from its fastener and unbolt the latch.

6. Check for any nameplates or emblems that are on the deck lid. Some are secured ☐
   with two-sided tape or small screws, bolts, or barrel clips on the backside. (Remember, even though you are **NOT** removing these on this exercise, you always want to
   be careful not to scratch the paint when removing any emblems.) There is a separate
   job sheet for removing and installing emblems and moldings.

   a. What is securing the emblems on your panel? _____

   _____

7. Position one partner on each side of the deck lid to hold the front edge of the panel. ☐
   What is the reason for doing this? _____

   _____

8. Remove the lid and place it face up on a covered workbench or stand. ☐

Reinstallation

9. With the help of your two partners, reinstall the deck lid, aligning the hinge flanges ☐
   with the pencil marks you made during removal. Start all the bolts first and then
   hand-tighten them. Then tighten only the rear bolts on each side. You only need to
   secure it enough to check the alignment on your gaps.

10. Slowly close the lid and check all your gaps. At this point the latch should still be ☐
    out, so the lid should close with no effort.

    a. Do all your gaps match your original measurements? _____

    _____

    b. If not, keep realigning until they do.

**Task
Completed**

11. Now reinstall the latch; install the rubber grommets on the edge of the lid if you   ☐
    removed them already.

    a. When the trunk is closed, does the latch pull the deck lid to one side? _____

    _____

    b. If so, loosen the latch bolts and realign until it pulls straight down and easily
        latches.

12. Once the latch is adjusted properly, adjust one of the end rubber grommets all the   ☐
    way out if it has them. If it does not, disregard 4a and 5a.

    a. How does this affect the height of the deck lid? _____

13. Adjust one of the grommets all the way in.   ☐

    a. How does this affect the height of the deck lid? _____

    _____

14. Check for water, dust, and air leaks. Have your partner get inside the trunk with a   ☐
    flashlight. Close the lid. Run water along the perimeter of the lid, starting at the
    bottom and working your way up.

    a. How long should water be sprayed around the deck lid to find the leak? _____

    _____

    b. Were any leaks detected using water? _____

    c. Can your lid be checked using a light? _____

    d. Were any leaks detected using a light? _____

INSTRUCTOR'S COMMENTS _____

_____

_____

Name _____ Date _____

Remove and Install Fender, Bumper, Headlight, and Grille

Objective

Upon completion of this activity sheet, the student should be able to inspect, remove, replace, and realign any fender, bumper, bumper cover, reinforcements, headlights/taillights, grilles, and any other related components.

ASE Education Foundation Task Correlation

II.A.1	Select and use proper personal safety equipment; take necessary precautions with hazardous operations and materials in accordance with federal, state, and local regulations. **(HP-I)**
II.A.2	Locate procedures and precautions that may apply to the vehicle being repaired. **(HP-I)**
II.A.3	Identify vehicle system hazard types (supplemental restraint system [SRS], hybrid/electric/alternative fuel vehicles), locations, and recommended procedures before inspecting or replacing components. **(HP-I)**
II.B.8	Prepare damaged area using water-based and solvent-based cleaners. **(HP-I)**
II.B.10	Inspect, remove, and reinstall repairable plastics and other components for off-vehicle repair. **(HP-I)**
II.C.1	Inspect/locate direct, indirect, or hidden damage and direction of impact. **(HP-I)**
II.C.2	Inspect, remove, and replace mechanically fastened welded steel panel or panel assemblies. **(HP-G)**
II.C.8	Inspect, remove, replace, and align bumpers, covers, reinforcements, guards, impact absorbers, and mounting hardware. **(HP-I)**
II.C.9	Inspect, remove, replace, and align fenders and related panels. **(HP-I)**
III.C.9	Check operation and aim headlight assemblies and fog/driving lights; determine needed repairs. **(HP-I)**

We Support

Education Foundation

Tools and Equipment

Vehicle
Creeper
Bumper jack or floor jack
Tape measure
Flashlight
Appropriate screwdrivers, ratchets,
    sockets, and specialty clip removal tools

Safety Equipment

Safety glasses or goggles

Introduction

Most of the time on modern vehicles, to remove bumper covers, you will almost always have to remove a grille that will either be separate or attached to the bumper cover. Also, the headlights must be removed in some cases before removing the bumper, sometimes not until after the bumper is removed. After the necessary components are removed and have access to the fender, you will then need to remove the fender. It is a very situational task that varies from one vehicle to another. In any case, to complete this job sheet, you must remove, reinstall, and realign all the components.

Vehicle Description

Year _____ Make _____ Model _____

VIN _____

NOTE: Before starting the removal process, a quick internet or YouTube search for OEM-specific instructions can be done to assist in removal instructions.

Grille Removal (if it is separate from the bumper)

Task Completed

This is a very situational process. Sometimes the grille will be attached to the bumper and will be removed with the bumper. If this is the case, omit this section; it is covered later in this exercise.

1. Locate the bolts or pushpins on top of the grille shell. Look all around the perimeter (outer edges) of the grille shell for any more visible fasteners holding it in place.   ☐

2. Remove those fasteners. If the grille does not easily lift out, there may be more hidden fasteners holding it in place. With a flashlight look down through any openings and locate any other hidden retainers, fasteners, or bolts holding it in place. Remove any hidden fasteners and the grille shell.   ☐

Headlight/Taillight Removal

This is also a very situational process. Sometimes the headlights/taillights (depending on front or rear bumper) need to be removed before the bumper, so you can access hidden fasteners underneath to remove the bumper. Sometimes the bumper must be removed to access the hidden fasteners at the bottom of the fender where it meets the bumper cover, and go through the fender bracket into the headlights. It all depends on the manufacturer's design. Your responsibility is to determine which method must be used.

1. Locate the headlight wiring connectors and disconnect them.   ☐

   a. Did you break any of the connectors?

   _____

2. Locate the fasteners securing the headlights.   ☐

3. Remove the fasteners and properly label and store them in a Ziploc bag.   ☐

   a. Did you break any fasteners? If so, list which one and how many.

   _____

   _____

Bumper Removal (Trucks and some older SUV models. If this vehicle is a car, omit this section.)

Before starting the removal process, make sure you have your safety equipment on.

1. Are you removing the front or rear bumper? _____

2. Stoop down and look directly at the center of the bumper from about 5 feet away to check for alignment differences or unevenness.   ☐

**Task
Completed**

3. Using a tape measure, measure the gap between the bumper and the fender on
    each side of the vehicle **ONLY** if it is a truck or an SUV. If it is a car, no gap should
    be present as they fit up flush.   ☐

    a. What are the gap measurements from both sides of the bumper?

    _____

4. Tape the edges of the fenders or quarter panels with strips of a 2-inch tape.   ☐

    a. What is the reason for doing this? _____

5. It is always best to work with a partner, but bumpers can be safely removed by one
    person. Using a floor jack or a bumper jack, lift on the center of the bumper until
    slight pressure is applied. You may have to remove a lower valance panel to secure
    the jack under the steel bumper itself. Never jack a bumper up by the plastic. Make
    sure you use a rag or towel to cover the jack, so no damage occurs to the bumper.   ☐

    a. Did any damage occur when removing the bumper?

    _____

6. Many bumpers have fog lights in them. Check to make sure both fog lights
    function, and then disconnect them before removing the bumper so you do not
    damage the wiring.   ☐

    a. Did all the lights work properly before removing?

    _____

7. While lying under the vehicle, carefully disconnect the wiring connectors from the
    fog lights.   ☐

8. It may be necessary to refer to the factory service manual or Google/YouTube to
    locate any other parts that may be holding the bumper in place. Remove the bum-
    per and place on a clean workbench.   ☐

Bumper Cover Removal (Cars and newer SUV models. If this vehicle is a truck, omit this section.)

Most of the time bumper covers are fastened to the fenders with small bolts in the upper
corners near the wheel, retaining pushpin clips, bolts, and/or screws underneath to
the lower splash shield, mud guards, and sometimes in the inner lip where there will be
pushpin clips fastening it to the inner splash shield along with other locations. Locate
these fasteners and remove them. Make sure to put all fasteners into Ziploc bags and
label each bag properly.

1. Did you break any fasteners that mount the bumper cover? If so, list how many and
    what kind. _____

2   With another person to help you, carefully remove the bumper cover from the
    vehicle and place on a padded work stand.   ☐

Grille Removal and Reinstallation (if fastened to the bumper)

1. If the grille is fastened inside the bumper cover, now you will be able to have access
    to the grille fasteners. Properly store them in a Ziploc bag.   ☐

    a. Once removed, get your instructor to show that you have properly removed the
        grille and labeled everything correctly.   ☐

    b. Did you break any fasteners? _____

Fender Removal

Once the necessary components are removed and you have access to the fender, you will then need to remove the fender. The most common places for the fasteners are as follows: There will almost always be bolts that are along the top (inside the hood-opening area), at the front lower area where the bumper cover snaps into place, along with push in retainers and/or screws holding the inner fender and mud flap to the fender, inside the door opening (after opening the door for access), and at the lower rear portion of the fender (sometimes you have to remove a rocker molding to gain access to these bolts).

1. Remove all necessary hardware and fasteners, and then place the fender on a work stand with the painted side facing up so that it does not get scratched or damaged. ☐

2. Did you break any fasteners? If so, how many and what kind?
   _____

3. Go get your instructor to show that you have removed everything and labeled everything properly. ☐

4. You may now reinstall all the components that were removed, in reverse order of the way you removed them. ☐

Fender Reinstallation

You may now reinstall the fender, placing it into position on the vehicle. Since this is your starting point of reinstalling all of the components in this job sheet, it is critical that this be done correctly. If not, all of the following components will not be aligned or fit together properly.

1. Hand-start all the bolts, making sure to leave them all loose so that you can realign the fender properly. Once properly aligned to the door and hood, you may then tighten all the bolts down. Make absolutely sure the fender is aligned correctly. ☐

2. Get your instructor's approval before continuing to the next step. ☐

3. Reinstall the fender liner and/or mud flap to the fender. ☐

4. Reinstall the rocker molding (only if it had to be removed for access). ☐

Bumper Reinstallation (For truck/SUV only. If this is a car, omit this section.)

1. For reinstallation of a car bumper, reverse the instructions of the removal process. ☐

2. Why is this important to hand-start all the bolts before you tighten any of them down?
   _____

   a. Is the bumper centered and level? _____.

      If not, make necessary adjustments until it is.

**Task
Completed**

3. Measure the distance between the bottom of the fenders to the top of the bumper.   ☐

   a. Are these measurements the same as your original measurements? If they are not, explain why. _____

_____

4. Reinstall the fog lights and the lower valance. Fill in the following checklist to make sure that all the lights are working.   ☐

   a. Do the fog lights work the same as they did before they were removed?

_____.

     If not, check the wiring connectors for any loose connections.

5. Wash and clean any contaminants or fingerprints off the area with the appropriate cleaners.   ☐

Bumper Installation (Car only. If truck, refer to preceding instructions.)

1. For reinstallation of a car bumper, reverse the instructions of the removal process.   ☐

   a. Does the bumper sit flush to the adjacent panels? _____.

     If not, adjust accordingly until they do.

   b. Do the fog lights work the same as they did before they were removed?

     _____. If not, check the wiring connectors for any loose connections.

2. Check for any missing or broken fasteners that need to be replaced.   ☐

   a. Do you have any extra fasteners left over? _____

     If so, find the locations that are missing fasteners and reinstall them.

Headlight Aim and Adjustment

Always make sure to check the aim of the headlights before returning a vehicle to the customer. You should see adjuster screws that need to be either turned clockwise or counterclockwise to adjust. Some headlights have left-right and up-down adjusters; some only have up-down depending on the make and model.

1. Make sure the area to be used is on a level surface.   ☐

2. Park the vehicle up close to the wall, and turn the headlights on low beam.   ☐

3. Make sure front wheels are pointing straight.   ☐

4. Mark the locations of where the beams hit the wall with a piece of 2-inch masking tape.   ☐

5. Now back up the vehicle 25 feet on the level surface. Turn the headlights on, and see if they align with the masking tape locations on your wall.   ☐

6. Next, find the adjusters, which are often located somewhere on the headlight housing. Each make and model is different, but generally the adjusters are a type of screw or bolt on the back and side of the headlight unit. While they are not often marked, they are often gray or silver, which stands out from the black headlight backs.   ☐

7. Refer to each individual manufacturer about the acceptable distance between where the lamps are shining 25 feet away and the original tape line.   ☐

**Task
Completed**

8. Many automakers offer headlight-aiming specs. Most say within a 2-inch distance   ☐
    is acceptable. A look at the owner's manual or a quick Google search can help.

9. Keep adjusting until you get the aim within the acceptable tolerance.   ☐

INSTRUCTOR'S COMMENTS _____

_____

_____

Review Questions

Name _____ Date _____ Instructor Review _____

1. The measurable distance between body panels is called a
   _____.

2. A _____ is the panel over the front tire that covers the inside of the fender.

3. _____ allow the hood to open and close.

4. Bumpers are designed to withstand major impact.
   A. True
   B. False

5. If a foam energy absorber is damaged in an accident, it can be repaired.
   A. True
   B. False

6. You should always test fit a part before it is painted or permanently installed.
   A. True
   B. False

7. A hood is not closing properly. Technician A says that the hood alignment may need to be adjusted. Technician B says that the latch may need to be adjusted. Who is correct?
   A. Technician A
   B. Technician B
   C. Both Technician A and Technician B
   D. Neither Technician A nor Technician B

8. Technician A says that a heat gun is best to use to soften the adhesive on moldings. Technician B says that a torch with a very low flame is better. Who is correct?
   A. Technician A
   B. Technician B
   C. Both Technician A and Technician B
   D. Neither Technician A nor Technician B

9. Technician A says that some hood latches are adjustable. Technician B says that all hood latches are adjustable. Who is correct?
   A. Technician A
   B. Technician B
   C. Both Technician A and Technician B
   D. Neither Technician A nor Technician B

10. Technician A says that older metal bumpers provide more safety to passengers because of their strength. Technician B says that modern bumpers have absorbers that help contain some of the impact, so they are safer for the passengers. Who is correct?
    A. Technician A
    B. Technician B
    C. Both Technician A and Technician B
    D. Neither Technician A nor Technician B

Door, Roof, and Moveable Glass Service

Name _____ Date _____ Instructor Review _____

Parts Definitions

Without using your textbook, name the door parts that are defined next.

1. Engages the striker on the body to hold the door closed: _____

_____

2. Fits around the door opening to seal the door to body joint: _____

_____

3. The main steel frame of the panel: _____

_____

4. Fits between the inner trim panel and frame to keep out moisture and wind noise: _____

_____

5. Allows for good visibility out of the door: _____

_____

6. The outer panel: _____

_____

7. Uses linkage rods to transfer motion to the door latch: _____

_____

8. Serves as a guide for the window to move up and down: _____

    _____

9. Attractive cover over the inner door frame: _____

    _____

10. A motor and a set of guides or tracks for moving the glass up and down: _____

    _____

Name _____ Date _____ Instructor Review _____

Door Parts Identification

Using the numbers on the following figure, name each part.

1. _____

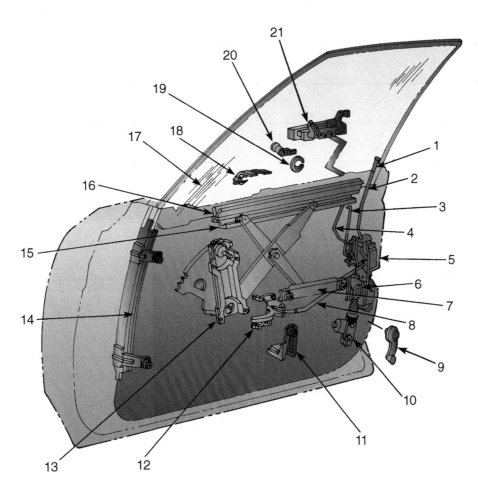

2. _____

3. _____

4. _____

5. _____

6. _____

7. _____

8. _____

9. _____

10. _____

11. _____

12. _____

13. _____

14. _____

15. _____

16. _____

17. _____

18. _____

19. _____

20. _____

21. _____

Name _____ Date _____

Remove and Install Door and Internal/ External Components

Objective

Upon completion of this activity sheet, the student should be able to inspect, remove, replace, and align a door and all related internal/external components (door panels, handles, belt line moldings, glass, and regulator).

ASE Education Foundation Task Correlation

II.B.2	Inspect, remove, label, store, and reinstall exterior trim and moldings. **(HP-I)**
II.B.3	Inspect, remove, label, store, and reinstall interior trim and components. **(HP-I)**
II.B.6	Protect panels, glass, interior parts, and other vehicles adjacent to the repair area. **(HP-I)**
II.C.6	Inspect, remove, replace, and align doors, latches, hinges, and related hardware. **(HP-I)**
II.C.14	Diagnose and repair water leaks, dust leaks, and wind noise. **(HP-G)**
II.E.1	Inspect, adjust, repair, or replace window regulators, run channels, glass, power mechanisms, and related controls. **(HP-I)**
II.E.2	Inspect, adjust, repair, remove, reinstall, or replace weatherstripping. **(HP-G)**
II.E.5	Initialize electrical components as needed. **(HP-G)**

We Support
ASE | **Education Foundation**

Tools and Equipment

Vehicle
Floor jacks
Appropriate screwdrivers, ratchets, and sockets
Battery service tools
Specialty door removal tools
Masking tape

Safety Equipment

Safety glasses or goggles
Work gloves

Introduction

Doors are complex and heavy assemblies that must protect the driver and passengers during a collision. Doors often have to be removed and replaced during collision repairs. This job sheet will help you learn how to safely remove, install, and adjust a vehicle's door and remove and install all related components of a door.

Vehicle Description

Year _____ Make _____ Model _____

VIN _____

Procedure

Door Remove and Install

**Task
Completed**

1. Before starting any work, disconnect the battery. Make sure they do not touch each other or any other metal at the same time.   ☐

2. Before removing the door, check the opening and closing action of the door assembly.   ☐

   a.   When first opened, does the door sag? _____

3. Have a partner move the door up and down when halfway open while you use a droplight to look for wear and play in the door hinges. **NOTE:** If there are no pins and bushings, omit this section.   ☐

   a.   Is there any play in the pins and bushings?

   b.   Must the door be lifted to properly close the door?

4. To remove the inner door trim panel, the fasteners holding it in place are generally hidden for looks purposes. Always look for hidden screws or bolts under plastic pop-out plugs, inside the grab handle area, behind the inner door handle, or around the edges of the panel itself.   ☐

   NOTE: A quick internet or YouTube search can help with specific locations as this is a general guide.

   a.   If your door has a manual window, it is normally secured with a C-clip. Use the appropriate door handle remover tool to remove the C-clip.

   NOTE: Keep your hand or a magnet close to the handle to catch the C-clip, as it likes to fly across the room and become lost forever most times.

5. Next, if the power switch assembly is a separate piece from the door panel, it sometimes has to be removed to get to a bolt behind it. Gently get a nonmarring pry tool in between and lift up being careful not to break anything. Using a flashlight, check to see how it comes out. **NOTE:** You may want to try taking the door panel off first as sometimes it is screwed to the backside of the panel and it will break if you try to pry it off from the front.   ☐

6. If all known fasteners have been removed, slowly and carefully grab the door panel and lift up. If it does not slide up, then it should have push in clips holding it to the door frame. Starting in the lower corner closest to the hinges, pull the door panel slightly to pop it loose. **NOTE:** You may need a trim remover tool or a nylon nonmarring pry tool to help pop the panel loose. Using a creeper may also help to examine under the door with a droplight so you do not miss any hidden clips.   ☐

7. Place the inner panel face up on a clean workbench or on padded work stands. **DO NOT LAY THEM FACE DOWN ON THE CONCRETE.**   ☐

8. Next, remove the moisture barrier covering the inside of the door frame opening. Older models were more of a brown paper type material. Most modern models are clear plastic. Pull one corner of it slowly to avoid tearing it. Make sure to have gloves on so the black butyl tape does not get all over your hands.   ☐

   a.   What is the purpose of this layer of plastic over the door opening?

   _____

   _____

**Task
Completed**

9. Disconnect any wires running inside the door through the pillar. Some unplug   □
easily, whereas others can be difficult to disconnect. Refer to a service manual or
a quick Google or YouTube search for the specific vehicle.

10. Next, to remove the door, a door jack is the best and safest equipment to use. If   □
you do not have one, a floor jack will suffice. Place a floor jack under the bottom
of the door. Fit a notched door-supporting tool or a 2 × 4 covered with a rag into
the jack saddle. This will allow you to hold the weight of the door and keep it from
falling when you remove the door hinge bolts. Center the jack and tool in the
middle of the door so that it is balanced on the jack.

11. Raise the jack until the tool or 2 × 4 just touches the bottom of the door. You only   □
want light upward pressure on the door. **DO NOT LIFT THE DOOR ENOUGH TO
STRAIN THE DOOR HINGES.**

12. Remove the bolts holding the hinges onto the door. Do not remove the bolts   □
holding the hinge to the car body. **It is important that you DO NOT loosen the
hinge bolts on the pillar side as this will make adjusting much more difficult
when reinstalling the door.**

13. Before removing the last hinge bolt or pin, have your partner hold the door on the   □
jack to keep it from falling. With door hinge fasteners removed, place the door on
padded stands or a workbench covered with a fender cover. Place the outside of
the door facing down.

14. Check the rubber weatherstripping on the door. Inspect it for tears, splits, and   □
other problems.

  a.   How will its condition affect water or wind leaks around the door?

  _____

  _____

Removing an Outer Door Handle

15. Locate the bolts or fasteners holding the outer door handle in place. They can   □
be inside the door structure and may require using a flashlight to find. If this is
your design, locate and remove the bolts. Using a flashlight, look inside the door
structure and find the connecting rods or cables running to the outer door handle
and disconnect them. **NOTE**: Be careful to not break any clips in the process.

16. Some models have a black plastic cap between the outer door skin and the   □
weatherstripping that will have to be removed for access to the securing bolt.
Once that bolt is removed or loosened, the outer handle bezel (small piece
next to the door handle) can be pulled out. Then by grabbing the door handle
and sliding it toward the opening where the bezel was, you should be able to
remove the door handle. **Make sure to disconnect any wiring connectors if
applicable.**

Removing an Outer Mirror

17. Disconnect the wiring harness connector from the mirror. Locate the bolts hold-   □
ing the mirror into place. Most times there are three bolts. After taking the first
two out, have a partner hold the mirror while you remove the third bolt. Then
remove the mirror.

Removing an Outer Belt Molding (inner belt molding should just lift off easily in most cases)

Task Completed

18. Locate any trim screws on the inside edge of the door that would be securing the belt molding to the door and remove them. **NOTE: Remember, the mirror must be removed on front doors before the belt molding can be removed or it will be damaged.**   ☐

19. Gently lift up to see if it will come off easily. If not, get a belt molding remover tool and remove the belt molding being careful not to distort or damage it in any way.   ☐

Removing a Window Regulator, Glass, and Run Channel

20. Lower the window until the regulator-to-glass bolts are visible. Remove these bolts.   ☐

21. Sometimes the door glass is riveted to the regulator. If this is the case for your door, you must knock out the center pin with a hammer and a punch. Then drill the head off the rivet with a slightly larger size drill bit than the opening of the rivet head.   ☐

    a.   How was the glass secured to the regulator?

    _____

22. Gently lift the glass out of the door shell, and place it on a workbench covered with a blanket. Sometimes the window run channel must be removed either partially by removing one side or all the way to get the glass to come out. If the glass does not come out easily, then remove the run channel until it slides out.   ☐

23. Now remove the bolts securing the regulator to the door shell. Remove the regulator through the largest opening in the door.   ☐

24. Before reinstallation begins, check inside the bottom of the door for any parts that may have fallen. It is also a good idea to vacuum out any dirt or rust that may be on the inside bottom of the door. Remember that quality work along with strong customer satisfaction will ensure repeat business. A door full of rust and loose parts on the bottom can rattle and annoy the owner.   ☐

25. Now get your instructor to check your work before you start the reassembly process.   ☐

Reinstalling All Components and Hardware (handles, belt moldings, glass, regulator, and run channels)

26. Reinstall by the same steps as removal in reverse order listed earlier. Remember the order in which everything came out. First item taken off is the last to go back on.   ☐

27. Make sure all run channels, regulator, and door glass weatherstripping are installed correctly by visually inspecting and checking each bolt and/or screw.   ☐

28. Check the operation of the window by rolling it up and down all the way.   ☐

    a.   Does the window roll all the way up? _____. If not, loosen the fasteners and adjust until it does.

    b.   Are there any gaps in either corner of the window that could cause an air or a water leak? _____

29. Check the operation of the inner/outer door handles (and mirror if applicable) to be sure they function properly.   ☐

	Task Completed

Reinstalling a Door

30. Place strips of a 2-inch-wide masking tape on all adjacent panels of the door that could be hit and damaged during door installation. ☐

31. Adjust the height of the floor jack so that it is even with the top of the rocker panel. With the help of your partner, slowly and carefully place the door on the jack. The weight of the door must be centered or balanced on the jack. The door should be in the open position about 3 to 4 inches from the hinges on the pillar post. If necessary, feed any wire loom hanging from the post into the door. ☐

32. You may use a small punch to line up the holes in the door to assist in alignment. ☐

33. Hand-tighten the hinge bolts only. When these are in, install the remaining bolts and tighten one bolt on each hinge. ☐

34. **To avoid body damage, never slam a door close right after installation.** Slowly close the door while making sure it does not jam into the fender or car body anywhere. If it starts to hit, loosen the bolts so you can raise or lower the hinges so that the door closes properly. You must close it easily to check that it is adjusted properly. ☐

35. Make sure that all body lines match up and all gaps are flush. If they are not, keep adjusting until properly aligned. When all the gaps are aligned, tighten the remaining hinge bolts. ☐

36. With the door secured and aligned, reinstall any components that have not been replaced until the door is fully reassembled and operational. ☐

37. If applicable, reinitialize any pre-set functions that may have been associated with the door panel (mirror settings, sometimes seat settings, etc.). ☐

38. When this is completed, get your instructor to check your work. ☐

Perform Air/Water Leak Tests

39. Perform an air leak by rolling the window all the way up. Have a partner take an air blower and turn on the outside of the glass at the top. Put your hand on the inside and feel for air. ☐

   a. Was there a leak? _____. If so, adjust the glass until there is no leak.

40. Perform a water leak test by rolling the window all the way up. Using a water hose at light pressure at the top of the glass seal area, see if any water runs in the seal of the glass. A spray bottle can be used, but sometimes it is not as effective as it is not a constant stream of water. ☐

   a. Was there a leak? _____. If so, remove the door panel and adjust the glass until there is no leak.

INSTRUCTOR'S COMMENTS _____

_____

_____

Name _____ Date _____

Door Skin Replacement

Objective

Upon completion of this activity sheet, the student should be able to safely and accurately replace a welded-on door skin.

ASE Education Foundation Task Correlation

II.A.4	Select and use a NIOSH-approved air purifying respirator. Inspect condition and ensure fit and operation. Perform proper maintenance in accordance with OSHA regulation 1910.134 and applicable state and local regulations. **(HP-I)**
II.B.3	Inspect, remove, label, store, and reinstall interior trim and components. **(HP-I)**
II.B.4	Inspect, remove, label, store, and reinstall body panels and components that may interfere with or be damaged during repair. **(HP-I)**
II.B.5	Inspect, remove, protect, label, store, and reinstall vehicle mechanical and electrical components that may interfere with or be damaged during repair. **(HP-G)**
II.B.6	Protect panels, glass, interior parts, and other vehicles adjacent to the repair area. **(HP-I)**
II.C.2	Inspect, remove, and replace mechanically fastened welded steel panel or panel assemblies. **(HP-G)**
II.C.6	Inspect, remove, replace, and align doors, latches, hinges, and related hardware. **(HP-I)**
II.C.10	Restore corrosion protection during and after the repair. **(HP-I)**
II.C.11	Replace door skins. **(HP-G)**
II.C.12	Restore sound deadeners and foam materials. **(HP-G)**
II.C.13	Perform panel bonding and weld bonding. **(HP-G)**
II.C.16	Weld damaged or torn steel body panels; repair broken welds. **(HP-G)**
VI.A.1	Select and use proper personal safety equipment; take necessary precautions with hazardous operations and materials in accordance with federal, state, and local regulations. **(HP-I)**
VI.A.2	Locate procedures and precautions that may apply to the vehicle being repaired. **(HP-I)**
VI.A.3	Identify vehicle system hazard types (supplemental restraint system [SRS], hybrid/electric/alternative fuel vehicles), locations, and recommended procedures before inspecting or replacing components. **(HP-I)**
VI.A.4	Select and use a NIOSH-approved air purifying respirator. Inspect condition and ensure fit and operation. Perform proper maintenance in accordance with OSHA regulation 1910.134 and applicable state and local regulations. **(HP-I)**
VI.B.1	Identify the considerations for cutting, removing, and welding various types of steel, aluminum, and other metals. **(HP-G)**
VI.B.2	Determine the correct GMAW welder type, electrode/wire type, diameter, and gas to be used in a specific welding situation. **(HP-I)**
VI.B.3	Set up, attach work clamp (ground), and adjust the GMAW welder to "tune" for proper electrode stick-out, voltage, polarity, flow rate, and wire-feed speed required for the substrate being welded. **(HP-I)**

VI.B.7   Identify hazards; foam coatings and flammable materials prior to welding/cutting procedures. **(HP-G)**

VI.B.11   Determine the type of weld (continuous, stitch weld, plug, etc.) for each specific welding operation. **(HP-I)**

VI.B.16   Identify cutting process for different substrates and locations; perform cutting operation. **(HP-I)**

We Support

ASE | Education Foundation

Tools and Equipment

Door in need of skin replacement
Door skin bonding adhesive
Hammer and dolly
MIG welder
Flange tool
Metal file
DA
Grinder
Belt sander
Assorted vise grips

Safety Equipment

Safety glasses or goggles
Ear protection
Dust respirator
Work gloves
Welding gloves
Welding jacket or sleeves
Welding respirator
Welding helmet

Introduction

Door skins are often replaced. This is sometimes cheaper than replacing the complete door, yet it still gives the customer a new panel. Proper measuring is crucial so that the door lines up after skin installation.

If you have any difficulty or doubt about any aspect of this activity, go to 3Mcollision.com and click on applications. Then select Metal Shop. Scroll down to the "Door Skin Removal" and the "Door Skin Replacement" videos and study them. After watching these you may ask your instructor for help or to answer your questions if needed.

Vehicle Description

Year _____ Make _____ Model _____

VIN _____

Procedure

Task Completed

1. Remove the trim pad, glass, and any handles, and disconnect any electrical wires on the inside of the door. With the help of a partner, remove the door.   ☐

    a.   Was the battery disconnected before removing the door wires? _____

    b.   What is the importance of disconnecting the battery? _____

    _____

    c.   Are the door hinges welded or bolted on? _____

**Task
Completed**

2. With a grinder, grind around the edge of the skin until you see the outer skin edge   ☐
   separate into two. Once the edges separate, stop grinding. Do not grind on the
   edge of the door shell itself. Clamp a pair of vise grips to the hem flange to help
   remove the flange as you are grinding.
   a.   Why is it important to not grind on the inner door shell?

   _____

3. Cut off the skin on the window frame as close to the door as possible.   ☐

   a.   Did you scar the window frame with the grinder at all? _____

   b.   If so, how are you going to correct this? _____

4. Cut off any tabs on the inside of the window opening. Be careful not to force any   ☐
   parts, because they could easily bend, making alignment of the new panel more
   difficult. If necessary, apply heat to loosen panel adhesive that may be bonding
   the skin to the door frame.

5. Remove the skin. Be careful of any sharp edges.   ☐

   a.   What is the condition of the inner door shell? _____

   _____

6. With the help of your partner, flip the door over.

   a.   Are there any spot welds holding pieces of the old skin onto the shell?

   _____

7. If there are still small pieces of the spot weld remaining, grind them down level to   ☐
   the door shell with a belt sander.

   a.   Why is it important to have a level and smooth edge when installing the new
        door skin? _____

   _____

8. Determine the window frame splice location, and flange the window frame panel.   ☐
   Prep the tabs for weld by grinding or sanding them down to bare metal. If you are
   replacing the full door skin, disregard this question.

9. Test the fit of the new skin. Line up the body lines. When lined up, install one   ☐
   sheet metal screw on each end or vise grip into place if possible without the vise
   grips hitting the adjacent panels when the door will latch into place. With your
   partner's help, temporarily install the door back on the vehicle.

   a.   Why is it important to test fit the door skin before you go any further? _____

   _____

   b.   Are all the body lines straight? _____

   c.   If not, how will you adjust them? _____

   _____

   d.   If necessary, add sound deadener.

**Task
Completed**

10. Remove the door and the skin. Apply the primer and urethane bond as per kit instructions.   ☐

   a.   What brand of adhesive are you using? _____

   b.   What is the work time for the adhesive? _____

   c.   What is the cure time (can also be called sand time or paint time depending on manufacturer) for the adhesive? _____

11. Install the skin with proper alignment, using your screw holes. Fold over the hem flange. The most important step is to always support the door edge with a dolly; a small 2 × 4 or sometimes a small hard-surfaced rubber sanding block can be used as well. Start at one end and gradually work your way around the door. Do not try to fold over the lip of the skin in one pass. This can cause warpage. Work from one edge to the other, making sure you only hit on the 2 × 4, dolly, or hard-surfaced sanding block.   ☐

   a.   What is one advantage of using the 2 × 4 or hard-surfaced rubber sanding block instead of a dolly? _____

   _____

12. Before welding, check your welder. If this door skin requires no welding, leave the welding section blank.   ☐

   a.   What should the heat setting be turned to? _____

13. Use a metal file to locate high spots on the outside edge of the door.   ☐

   a.   Is there any damage from the hammer and dolly on the outside edge?

   _____

   b.   If so, how will you repair this damage? _____

   _____

14. Use #80 grit paper on a DA sander (an 8-inch sander also can be used for faster leveling). Smooth over the outside edges of the door with #180 grit and then with #320 grit. Check the door for high and low areas.   ☐

   a.   Why would you use a DA instead of a grinder, when the grinder would clearly be faster? _____

   _____

15. The repair is now done and ready for primer. That will be done on a separate job sheet when you get to the refinishing course.   ☐

INSTRUCTOR'S COMMENTS _____

_____

_____

Review Questions

Name _____ Date _____ Instructor Review _____

1. The _____ keeps the door from opening too wide.

2. A moisture barrier can be repaired with duct tape.
   A. True
   B. False

3. The _____ is the outer panel over the door frame.

4. Window regulators are always fastened with nuts and bolts.
   A. True
   B. False

5. If an intrusion beam gets bent, the entire door should be replaced.
   A. True
   B. False

6. Some door skins are secured with adhesives.
   A. True
   B. False

7. When replacing a door skin, Technician A says it is not necessary to test fit a new door skin because it is a new part. Technician B says pre-fit is always a good idea before installing a door skin. Who is correct?
   A. Technician A
   B. Technician B
   C. Both Technician A and Technician B
   D. Neither Technician A nor Technician B

8. A new front door has been installed and sticks out past the fender. Technician A says this is called negative alignment. Technician B says this will cause wind noise. Who is correct?
   A. Technician A
   B. Technician B
   C. Both Technician A and Technician B
   D. Neither Technician A nor Technician B

9. A sunroof is leaking water. Technician A says that the glass could need to be realigned. Technician B says that the drain tubes could be stopped up. Who is correct?
   A. Technician A
   B. Technician B
   C. Both Technician A and Technician B
   D. Neither Technician A nor Technician B

10. When removing broken glass from a run channel, Technician A uses a screwdriver. Technician B uses a plastic glass stick. Who is correct?
    A. Technician A
    B. Technician B
    C. Both Technician A and Technician B
    D. Neither Technician A nor Technician B

Repairing Sheet Metal

Name _____ Date _____ Instructor Review _____

Metal Straightening Terms

Explain the importance of each term in metal straightening.

1. Annealed _____

2. MS _____

3. HSS _____

4. Deformation _____

5. Tensile strength _____

6. Compressive strength _____

7. Shear strength _____

8. Torsional strength _____

9. Yield point _____

10. Elastic deformation _____

11. Spring back _____

12. Plastic deformation _____

13. Work hardening _____

14. Dinging hammers _____

15. Hammer on _____

16. Hammer off _____

17. Spoons _____

18. Stud welder _____

19. Heat shrinking _____

20. Paintless dent repair _____

Name _____ Date _____

One-Sided Dent Repair

Objective

Upon completion of this activity sheet, the student should be able to safely and accurately make a one-sided repair on any vehicle.

ASE Education Foundation Task Correlation

II.A.1	Select and use proper personal safety equipment; take necessary precautions with hazardous operations and materials in accordance with federal, state, and local regulations. **(HP-I)**
II.A.2	Locate procedures and precautions that may apply to the vehicle being repaired. **(HP-I)**
II.B.1	Review damage report and analyze damage to determine appropriate methods for overall repair; develop and document a repair plan. **(HP-I)**
II.B.9	Remove corrosion protection, undercoating, sealers, and other protective coatings as necessary to perform repairs. **(HP-I)**
II.C.1	Inspect/locate direct, indirect, or hidden damage and direction of impact. **(HP-I)**
II.C.3	Determine the extent of damage to aluminum body panels; repair or replace. **(HP-G)**
II.C.10	Restore corrosion protection during and after the repair. **(HP-I)**
II.D.2	Locate and repair surface irregularities on a damaged body panel using power tools, hand tools, and weld-on pulling attachments. **(HP-I)**
II.D.3	Demonstrate hammer and dolly techniques. **(HP-I)**
II.D.4	Heat shrink stretched panel areas to proper contour. **(HP-G)**
II.D.5	Cold shrink stretched panel areas to proper contour. **(HP-I)**
II.D.11	Straighten contours of damaged panels to a suitable condition for body fillings or metal finishing using power tools, hand tools, and weld-on pulling attachments. **(HP-I)**

We Support
ASE | **Education Foundation**

Tools and Equipment

Damaged vehicle
Stud gun set with draw pins and slide hammer
Electric dent puller with attachments
#80 grit paper
Various sized sanding blocks
Pick hammer
Body filler
Angle grinder with an 80 grit roloc disc

Safety Equipment

Safety glasses or goggles
Dust respirator
Work gloves

Introduction

Basic metal straightening, along with the proper use of hand tools and sandpapers, is an important skill when learning collision repair. The quality of the finished work always depends on the initial work. This exercise will help you develop the skills necessary to become a master craftsman.

Vehicle Description

Year _____ Make _____ Model _____

VIN _____

Procedure

1. This repair will be made on the outside only, for example, a door dent of a steel panel. If possible, find a door shell to work on. Have your instructor put several types of dents for you to repair.   ☐

    a.   How can you tell if it is steel or aluminum? _____
    _____

2. With an angle grinder, grind the paint to bare metal about 3 or 4 inches past the actual dent.   ☐

    a.   Why is it important to grind past the dent? _____
    _____

    b.   If your grinder cannot get in the dent, you will have to use a wire wheel in a drill, the eraser wheel tool, roloc wheel disc, or sand the paint off by hand with #80 grit.

NOTE: Search the internet or YouTube for 3M Metal working; the video "Removing Creases and Dents in Metal Work Panels" will walk you through the steps. Or you can watch the video titled "The Dent Fix Maxi-Multiple Pull Resistance Welder."

3. When you have pulled on all of the pins, continue tapping on any remaining ridges or high spots. Check the panel with a straightedge and by hand. Remove any remaining high or low spots until the surface is level.   ☐

4. Depending on the dent, if it flexes or "tin cans," you may have to shrink the metal. There are cold-shrinking and heat-shrinking methods. Search the internet for dentfix.com/videos, and then search "DF-505 Maxi Dent Pulling Station" for shrinking instructions. There are also videos available on YouTube on heat shrinking.   ☐

5. Some shops are equipped with an electric dent puller/shrinking tool. If your shop has this equipment, it is the easiest way to shrink metal. When shrinking small areas, you can use a draw pin welder and other equipment as well. A mini butane torch can be used also if necessary. Oxyacetylene torches were used frequently; however, due to thinner metals and more high-strength steels being used, it is no longer recommended as it generates too much heat to perform a safe repair.   ☐

6. If you do not have a repair station such as this that is capable of shrinking a panel for you, you will need a shrinking hammer and a dolly for this.   ☐

    a.   Which method are you using for this exercise? _____
    _____

    b.   If you are using hammer and dolly method, use steps 7–9. If you are using repair station method, follow the instructions from the video.

7. Hammer and dolly the area back as flat as you can get with a regular body hammer and dolly.   ☐

8. *Kinking* is another term for how you cold shrink metal and involves using a
hammer and dolly to create pleats, or kinks, in the stretched area to shrink its
surface area. Use your shrinking hammer and dolly for this. Hold the dolly lightly
against the back of the panel, and strike the front side of the panel with the
spiked surface of the shrinking hammer.   ☐

9. Use light taps or blows of the hammer around the outside of the stretched area
working your way inside and ending in the middle. Repeat until the metal does
not flex anymore. Be careful not to strike the hammer too much in the stretched
area, or it will deform the opposite way and shrink into a bulged area outward.   ☐

10. Low spots of less than ⅛ inch can be filled with body filler. Any low spot deeper
than ⅛ inch must be pulled out farther with the draw pins.   ☐

   a.   Why can dents of more than ⅛ inch not be filled? _____

   _____

11. Check the panel by hand to locate any remaining high or low spots. Repeat the
previous steps until there are no more high or low spots and ready for body filler.   ☐

12. Apply corrosion protection to the front and backside of the panel to treat any
areas that you may have broken through the paint or E-coat. Self-etching primer
may be used on the backside after smoothing out any scratched, chipped, or
burnt paint from a stud welder followed by an appropriate topcoat that matches
the original color coating of the original finish.   ☐

   a.   What did you spray on to treat the backside? _____

   b.   What did you spray on to treat the front side? _____

13. The application of body filler will be on another job sheet.   ☐

14. List here what steps would be different for this procedure if the panel were to be
aluminum.   ☐

   _____

   _____

   _____

INSTRUCTOR'S COMMENTS _____

_____

_____

Name _____ Date _____

Tension Panel Repair

Objective

Upon completion of this activity sheet, the student should be able to safely repair a quarter panel that has a tension dent or buckle.

ASE Education Foundation Task Correlation

I.B.1	Review damage report and analyze damage to determine appropriate methods for overall repair; develop and document a repair plan. **(HP-I)**
I.B.9	Remove corrosion protection, undercoating, sealers, and other protective coatings as necessary to perform repairs. **(HP-I)**
II.C.1	Inspect/locate direct, indirect, or hidden damage and direction of impact. **(HP-I)**
II.D.1	Prepare a panel for body filler by abrading or removing the coatings; featheredge and refine scratches before the application of body filler. **(HP-I)**
II.D.2	Locate and repair surface irregularities on a damaged body panel using power tools, hand tools, and weld-on pulling attachments. **(HP-I)**
II.D.3	Demonstrate hammer and dolly techniques. **(HP-I)**
II.D.4	Heat shrink stretched panel areas to proper contour. **(HP-G)**
II.D.5	Cold shrink stretched panel areas to proper contour. **(HP-I)**
II.D.9	Perform proper metal finishing techniques for aluminum. **(HP-G)**
II.D.11	Straighten contours of damaged panels to a suitable condition for body fillings or metal finishing using power tools, hand tools, and weld-on pulling attachments. **(HP-I)**

We Support

ASE | **Education Foundation**

Tools and Equipment

Vehicle with a damaged quarter panel
4- or 10-ton Porta Power
Various sized and shaped hammers
    and dollies
Straightedge
Heat source (butane torch, dent repair station,
    any means of applying minimal controlled
    heat for shrinking)
Draw pin welder

Safety Equipment

Safety glasses or goggles
Dust respirator
Work gloves

Introduction

Many times in the body shop, a work order will specify that a quarter panel should be replaced, even though a skilled technician could actually straighten and save the panel. When a quarter panel is severely damaged, it becomes shorter in length, making parts adjustment almost impossible. It is critical the panel be brought back to its original length to maintain original factory quality.

Vehicle Description

Year _____ Make _____ Model _____

VIN _____

Procedure

Task Completed

1. Obtain a quarter panel with a severe dent. ☐

2. Before starting operations, check the fluid level of the oil in the ram. Make sure it is full.

3. Set up a Porta Power inside the trunk. Set the ram against a bracket on the inner wheelhouse. If this is not stable, brace an area inside the trunk with pieces of 2 × 4 boards. Monitor this base during the push operation. ☐

   a.   Is your Porta Power a 4- or 10-ton unit? _____

4. Add extension tubes to the ram to reach the inner corner at the rear of the quarter panel. Your unit should have various heads to match the different contours you are trying to push. Remember, it is important to use the proper attachment head so the panel is not distorted. ☐

5. Now that you have the ram full of fluids, setup properly, and all safety equipment on, you may begin the procedure. ☐

6. Watch for movement in the lowest area of the dent. Make sure you have a partner to help watch for sudden damage. Apply only a small amount of tension to the dent. Follow the repair plan with a hammer and dolly, releasing the buckles in your planned sequence. Make sure you tap on the high spots while there is tension on the panel. Apply additional tension with the ram. Do not overextend the panel or it will stretch. ☐

7. Remember, pumping the ram should not take much effort. If you reach a point where you are forcing the pump, you can easily break the seal in the ram and cause an oil leak. If all the gaps are aligned, overextend slightly to allow for spring back. ☐

   a.   What happens to your panel when the tension is released from the ram?

   _____

8. Work any remaining high and low spots with a hammer and dolly. When an area pushes in and pops out, it is called oil canning. This must be repaired by shrinking. Some shops are equipped with an electric dent puller/shrinking tool. If your shop has this equipment, it is the easiest way to shrink metal. When shrinking small areas, you can use a draw pin welder and other equipment as well. A mini butane torch can be used also if necessary. Oxyacetylene torches were used frequently; however, due to thinner metals and more high-strength steels being used, it is no longer recommended as it generates too much heat to perform a safe repair. All oil canning must be removed before body filler can be applied. ☐

   a.   Is there any oil canning on your panel? _____

   b.   What process or method of straightening will you use to remove the oil canning? _____

   _____

**Task
Completed**

9. Check the panel for high areas by hand and using a straightedge. If no high areas are found and all low areas are less than ⅛-inch deep, the job sheet is now complete.

☐

   a. Do you have any high or low spots on your panel still? _____

10. Repeat the hammer and dolly process until all high or low spots are within ⅛-inch deep.

☐

The process of applying body filler and sanding will be on another separate job sheet.

INSTRUCTOR'S COMMENTS _____

_____

_____

Name _____ Date _____

Hood Repair

Objective

Upon completion of this activity sheet, the student should be able to safely repair a hood that has a tension buckle.

ASE Education Foundation Task Correlation

II.B.1	Review damage report and analyze damage to determine appropriate methods for overall repair; develop and document a repair plan. **(HP-I)**
II.B.9	Remove corrosion protection, undercoating, sealers, and other protective coatings as necessary to perform repairs. **(HP-I)**
II.C.1	Inspect/locate direct, indirect, or hidden damage and direction of impact. **(HP-I)**
II.C.3	Determine the extent of damage to aluminum body panels; repair or replace. **(HP-G)**
II.D.1	Prepare a panel for body filler by abrading or removing the coatings; featheredge and refine scratches before the application of body filler. **(HP-I)**
II.D.2	Locate and repair surface irregularities on a damaged body panel using power tools, hand tools, and weld-on pulling attachments. **(HP-I)**
II.D.3	Demonstrate hammer and dolly techniques. **(HP-I)**
II.D.4	Heat shrink stretched panel areas to proper contour. **(HP-G)**
II.D.5	Cold shrink stretched panel areas to proper contour. **(HP-I)**
II.D.9	Perform proper metal finishing techniques for aluminum. **(HP-G)**
II.D.11	Straighten contours of damaged panels to a suitable condition for body fillings or metal finishing using power tools, hand tools, and weld-on pulling attachments. **(HP-I)**

We Support

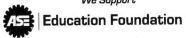 | Education Foundation

Tools and Equipment

Vehicle with a damaged hood
Appropriate body hammer and dollies
Wood block
DA sander
Spoon dolly
Draw pin welder/stud gun welder
Metal file
Tape measure

Safety Equipment

Safety glasses or goggles
Dust respirator
Work gloves

Introduction

The hood of a vehicle is probably the most replaced panel because the heat of the engine can cause filler work to rise and the hood is clearly visible to the driver. Often, the time it takes to straighten and refinish a hood is more expensive than replacing it. Even though in most cases the hood will be replaced, this job sheet will show you how to repair one in a very small case like this.

Vehicle Description

Year _____ Make _____ Model _____

VIN _____

Procedure

Task Completed

1. Make sure the vehicle you are working on has a buckle in or near the middle of the hood. This type of damage raises the hood above the level of the fender. ☐

   a. How much higher is the hood than the fender? _____

   b. What side of the hood is the damage on? _____

   c. How far back is the hood pushed? _____

   d. Why might the hood be pushed back if the damage is in the center of the hood? _____

2. Place a block of wood under the front edge of the hood. ☐

   a. What is the reason for this? _____

3. Use a dinging spoon and a ball peen hammer on the ridge. Hold the spoon tightly, and hammer rapidly on it. ☐

   a. What happens if the spoon is held loosely? _____
   If the ridge does not move, use a 2 × 4 board set on the edge in place of the spoon. The board should only be about 8–10 inches long.

   b. What is the length of the board you are using? _____
   Remove the block of wood, and check the hood for the level against the fender.

   c. What is the height of the hood at the fender now? _____

   d. How much lower is that than your original measurement? _____
   If the ridge is not in level with the fender, continue "spooning" the ridge. If access is available under the hood, the hammer and dolly can be used.

   e. What type of dolly will you use to best fit the underside contour of your hood? _____

4. Many hoods have a body line to stiffen the panel. A brick chisel can be used to lower the ridge of the body line. The chisel end of a hammer may also be used. ☐

   a. If the body line on your hood has a ridge in it, how long is the ridge?

   _____

   It is important to tap lightly when using tools, especially on a hood.

   b. What is the reason for this? _____

5. Remove the paint 4 inches on all sides using #80 grit on a DA sander. Do not use a grinder. ☐

   a. What is the reason for not using a grinder on a hood? _____

6. The metal around the ridge is softer than the ridge itself. Use a shrink fence to ☐
prevent this soft area from rising up during finishing. A shrink fence can be made
by surrounding the ridge with a series of pin shrinks. Use the draw pin welder to
tighten the metal adjacent to the ridge. Lightly tap the trigger of the welder to
create the smallest amount of heat.

   a.   Why is it best to use the smallest amount of heat? _____

   b.   How many pin shrinks did you use? _____
        Place a higher concentration of pin shrinks near the ridge.

   c.   What is the reason for this? Fewer shrinks are placed farther away from the
        ridge. An alternative shrink fence method is to use a sharp pick hammer
        instead of welds. _____

   d.   How can a pick hammer achieve the same results? _____
        _____

7.   Check the panel for high and low areas using a straightedge and by hand.   ☐
     Remove the remaining high spots with a draw pin welder, pick hammer, or shrink-
     ing hammer.

   a.   What is the difference between a pick hammer and a shrinking hammer?
        _____

8.   When all of the high areas are removed and all low areas are less than ⅛-inch   ☐
     deep with the panel tight, apply body filler. Sand the panel carefully as hoods are
     very easily distorted from even minimal heat.

   a.   What grit sandpaper did you use to sand the hood?
        _____

INSTRUCTOR'S COMMENTS _____

_____

_____

Review Questions

Name _____ Date _____ Instructor Review _____

1. Low-carbon steel is also called
   _____.

2. _____ are a result of bending metal past its elastic limit.

3. A _____ is a heavy steel block with various shapes on each side for straightening sheet metal.

4. It is okay to drill holes in a panel if you cannot access the backside for repairs.
   A. True
   B. False

5. If steel is cooled down suddenly, the steel contracts and its length is shortened.
   A. True
   B. False

6. Metal-working skills are probably the most important craft a body technician can bring to a shop.
   A. True
   B. False

7. Technician A says that paintless dent repair involves drilling holes in the outside of the panel. Technician B says that no holes are drilled in the panel. Who is correct?
   A. Technician A
   B. Technician B
   C. Both Technician A and Technician B
   D. Neither Technician A nor Technician B

8. Technician A says that a spoon can be used as a dolly. Technician B says that spoons can be used to pry out dents. Who is correct?
   A. Technician A
   B. Technician B
   C. Both Technician A and Technician B
   D. Neither Technician A nor Technician B

9. Technician A says that heat should not be applied to ultra-high-strength steel. Technician B says that heating this metal is acceptable as long as no holes are burned through. Who is correct?
   A. Technician A
   B. Technician B
   C. Both Technician A and Technician B
   D. Neither Technician A nor Technician B

10. Technician A says that damage such as a gouge, a tear, or a scratch is considered indirect damage. Technician B says that this type of damage would be considered direct damage. Who is correct?
    A. Technician A
    B. Technician B
    C. Both Technician A and Technician B
    D. Neither Technician A nor Technician B

Using Body Fillers

Name _____ Date _____ Instructor Review _____

Filler Application

In the chart that follows, write in the applications for each of the fillers without using your textbook.

Filler	Composition	Characteristics	Application
Conventional Fillers			
1 Heavyweight fillers	Polyester resins and talc particles	Smooth sanding; fine featheredging; nonsagging; less pinholing than lightweight fillers	
2 Lightweight fillers	Microsphere glass bubbles; fine grain talc; polyester resins	Spreads easily; nonshrinking; homogenous; nonsettling	
3 Premium fillers	Microspheres; talc; polyester resins; special chemical additives	Sands quickly and easily; creamy and moist; spreads smooth without pinholes; dries tack-free; will not sag	
Fiberglass Reinforced Fillers			
4 Short-strand	Small fiberglass strands; polyester resins	Waterproof; stronger than regular fillers	
5 Long-strand	Long fiberglass strands; polyester resins	Waterproof; stronger than short-strand fiberglass fillers; bridges small holes without matte or cloth	
Specialty Fibers			
6 Aluminum filler	Aluminum flakes and powders; polyester resins	Waterproof; spreads smoothly; high level of quality and durability	
7 Finishing filler/ polyester putty	High-resin content: fine talc particles; microsphere glass bubbles	Ultrasmooth and creamy; tack-free; nonshrinking; eliminates need for air dry-type glazing putty	
8 Sprayable filler/ polyester primer-surfacer	High-viscosity polyester resins; talc particles; liquid hardener	Virtually nonshrinking; prevents bleedthrough; eliminates primer/glazing/primer procedure	

Name _____ Date _____

Body Filler Exercise

Objective

Upon completion of this activity sheet, the student should be able to properly mix, apply, and finish body filler.

ASE Education Foundation Task Correlation

II.A.1	Select and use proper personal safety equipment; take necessary precautions with hazardous operations and materials in accordance with federal, state, and local regulations. **(HP-I)**
II.A.4	Select and use a NIOSH-approved air purifying respirator. Inspect condition and ensure fit and operation. Perform proper maintenance in accordance with OSHA regulation 1910.134 and applicable state and local regulations. **(HP-I)**
II.B.9	Remove corrosion protection, undercoating, sealers, and other protective coatings as necessary to perform repairs. **(HP-I)**
II.C.1	Inspect/locate direct, indirect, or hidden damage and direction of impact. **(HP-I)**
II.C.3	Determine the extent of damage to aluminum body panels; repair or replace. **(HP-G)**
II.D.1	Prepare a panel for body filler by abrading or removing the coatings; featheredge and refine scratches before the application of body filler. **(HP-I)**
II.D.2	Locate and repair surface irregularities on a damaged body panel using power tools, hand tools, and weld-on pulling attachments. **(HP-I)**
II.D.3	Demonstrate hammer and dolly techniques. **(HP-I)**
II.D.4	Heat shrink stretched panel areas to proper contour. **(HP-G)**
II.D.5	Cold shrink stretched panel areas to proper contour. **(HP-I)**
II.D.6	Identify body filler defects; correct the cause and condition (pinholing, ghosting, staining, over catalyzing, etc.). **(HP-I)**
II.D.7	Identify different types of body fillers. **(HP-G)**
II.D.8	Shape body filler to contour; finish sand. **(HP-I)**
II.D.10	Perform proper application of body filler to aluminum. **(HP-G)**
II.D.11	Straighten contours of damaged panels to a suitable condition for body fillings or metal finishing using power tools, hand tools, and weld-on pulling attachments. **(HP-I)**

We Support

Tools and Equipment

Body panel
Hand tools
Shrinking tools
Various sanding blocks
Various grits of sandpaper
Mixing board
Body filler

Safety Equipment

Safety glasses or goggles
Dust respirator
Work gloves

Spreader
Cheese grater
Straightedge

Introduction

Great time and money have been saved since the invention of body filler. This can be a very durable, long-lasting product if the repair is properly made. It is very important to have your repair area properly metal worked as body filler is only allowed to be ⅛-inch thick when final sanded. Any more than that and it is only hiding an improper repair and can crack out also.

Vehicle Description/Panel to Be Repaired

Year _____ Make _____ Model _____

VIN _____ or Type of Panel _____

Procedure

Task Completed

1. Obtain a medium crown panel. Use a ball peen hammer to make at least 10–20 small ☐
    dents in the metal area within a 4-inch × 12-inch area. Make sure to keep the area of
    dents small.

2. Grind the paint off the dented area with #80 grit, going 3 inches past the dent in all ☐
    directions.

    a. What is the reason for grinding farther than your work area? _____

    _____

    b. How will you get the paint out of the dents if the grinder will not reach? _____

    _____

    c. Why would you not continue to grind until you get the paint out of these dents?
        What would happen? _____

    _____

3. If the metal is stretched from excessive hammer force, you have hit the panel ☐
    extremely too hard and could require shrinking. If so, refer to the shrinking steps in
    the previous job sheets.

4. Lay a straightedge on the bare metal area. ☐

    a. How do you determine whether you have high spots? _____

    _____

    b. How are high spots removed? _____

    c. If high spots remain after filler application, what will happen when sanding?

    _____

    d. How can this be corrected? _____

5. Mix the body filler according to the manufacturer's recommendations. ☐

6. In the past, body filler was not designed to stick to paint. Now there are certain brands ☐
    and types of filler that will stick to properly sanded paint if the metal surface is just
    slightly dinged and does not require pulling. Some shops still want you to grind to
    the metal no matter what; some want you to sand the paint and apply filler directly
    to the paint as it keeps the original corrosion protection without grinding it off.
    You should be familiar with both procedures.

7. If grinding is required, grind paint back at least 3 inches. ☐

8. If the surface does not require pulling, sand the paint with #80 grit at least 3 inches ☐
beyond the dents. There should **NOT** be any shiny areas left to the painted surface
before the filler application. If there are still any shiny areas where you are going to
apply filler, repeat this step until the surface is dull.

   a. Why is it critical to remove all shiny spots from the area where filler will be applied?

   _____

9. Apply the body filler on the bare metal area or sanded area (whichever method you ☐
are using) with downward pressure on the surface. Use your spreader to work the filler
back and forth to make it smooth and even. Your filler should not be chunky and uneven
as this will create a surface area that is harder to work than necessary. The smoother
the filler spreads out, the easier it will be to shape and level the dent. Overfill the dent
so that you get the surface higher than it should be so you can sand it down level.

   a. What is the maximum thickness that body filler should be when final sanded?

   _____

   b. What does bare metal inside your repair area indicate when sanding the filler?

   _____

10. The most common mistakes are applying body filler before all of the high spots are ☐
removed. Avoid this problem by carefully checking the panel for high spots and
completely filling the low areas. If you encounter this problem, repeat the exercise
until you have accomplished it correctly.

11. If the surface was only a small dent, then a cheese grater may not be required. ☐
You may put #80 grit on a block and sand the top paraffin layer (sticky gummy layer) off.
Blow the gummed-up filler out of the paper before it hardens in the paper.

12. When the body filler has cured enough that it is hard but still tacky or gummy feeling, ☐
use a cheese grater (or technical name is a surform file) to quickly remove the excess
filler. If the body filler rolls off the bare metal, it has not cured enough. Allow additional
cure time. It should shred off like cheese being grated; hence, the nickname
cheese grater.

   a. How does temperature affect the curing time of filler? _____

   _____

13. When the body filler has been cheese-grated and has dried into a hard surface, sand it ☐
diagonally or in a crisscross X pattern. This will ensure you cover the largest possible
area. As you sand, keep the end of the long board on the undamaged areas.
This will allow you to match the contour of the panel.

14. The only time you should have to use #36 or #40 grit paper instead of #80 grit paper ☐
is to quickly knock down large areas of filler that have been applied and you forgot to
use a cheese grater. Most fillers today spread so smooth and thin that very coarse
paper is not as necessary as it once was. If your surface is aluminum, you cannot
sand it with any coarser than #80 grit anyway as it will shred the aluminum due to
it being a softer metal.

15. When the body filler becomes level, switch to #180 grit sandpaper on your sanding block.   ☐

    a. What is the advantage/disadvantage of using a sanding block over an air file?

    _____

16. Check the sanded area with your hand. The body filler must be smooth and level. Always try to run your hand horizontally along the panel. If you are having trouble feeling whether it is level or not, get a paper towel and put under your hand when you feel the surface. This will keep you from feeling the different textures of the metal and filler and will increase your sense of feel. Another trick to try is to close your eyes while running your hands over the surface. This takes away your sight and increases your sense of feel.   ☐

17. Have your instructor feel the repair. Do not get frustrated if your instructor feels high and low spots that you cannot feel. Your instructor has been in the industry for a long time and has developed a touch that can detect even slight imperfections.   ☐

18. Your goal is to do metal work, fill once, and finish the repair. A good standard to use is no more than two coats of filler to straighten a dent. If it takes more than two, your metal work is still either too high or low. Repeat this exercise until you can finish in two filler applications.   ☐

INSTRUCTOR'S COMMENTS _____

_____

_____

Review Questions

Name _____ Date _____ Instructor Review _____

1. Aluminum filler is used to _____
   _____.

2. Small pinholes and scratches can be filled
   with _____.

3. Guide coat is used to help find high spots in
   a repair area.
   A. True
   B. False

4. You must always shake the tube of hardener
   really well before mixing it with body filler.
   A. True
   B. False

5. Surface rust is an early stage of corrosion.
   A. True
   B. False

6. Wet sanding produces a very smooth surface
   that is free of scratches.
   A. True
   B. False

7. Technician A says that a guide coat is used
   to check for high and low spots. Technician B
   says that a guide coat is the last coat of color
   to be sprayed. Who is correct?
   A. Technician A
   B. Technician B
   C. Both Technician A and Technician B
   D. Neither Technician A nor Technician B

8. Technician A says it is okay to use cardboard
   to mix body filler. Technician B says body
   filler cannot be applied thicker than 1/8 of an
   inch. Who is correct?
   A. Technician A
   B. Technician B
   C. Both Technician A and Technician B
   D. Neither Technician A nor Technician B

9. Technician A says that featheredging is taper-
   ing the edges of the paint. Technician B says
   that featheredging is wet sanding primer.
   Who is correct?
   A. Technician A
   B. Technician B
   C. Both Technician A and Technician B
   D. Neither Technician A nor Technician B

10. Technician A says that a metal conditioner
    is used in place of a self-etching primer.
    Technician B says that a metal conditioner is
    another name for a prep solvent. Who is
    correct?
    A. Technician A
    B. Technician B
    C. Both Technician A and Technician B
    D. Neither Technician A nor Technician B

Repairing Plastics

Name _____ Date _____ Instructor Review _____

Plastic ID Lab

	Name	Thermoplastic/ Thermoset	Repair Method
1. PUR			
2. ABS			
3. PP			
4. TPO			
5. E/P			
6. UP			
7. SMC			
8. RRIM			
9. PPO			
10. PE			
11. PA			
12. ABS + PC			
13. PC			
14. TPR			
15. ABS + PVC			

Name _____ Date _____

Single-Sided and Two-Sided Plastic Adhesive Repairs

Objective

Upon completion of this activity sheet, the student should be able to safely and accurately repair scratches and tears in urethane bumpers.

ASE Education Foundation Task Correlation

II.A.1   Select and use proper personal safety equipment; take necessary precautions with hazardous operations and materials in accordance with federal, state, and local regulations. **(HP-I)**

II.A.4   Select and use a NIOSH-approved air purifying respirator. Inspect condition and ensure fit and operation. Perform proper maintenance in accordance with OSHA regulation 1910.134 and applicable state and local regulations. **(HP-I)**

II.F.1   Identify the types of plastics; determine repairability. **(HP-I)**

II.F.2   Clean and prepare the surface of plastic parts; identify the types of plastic repair procedures. **(HP-I)**

II.F.3   Repair rigid, semirigid, and flexible plastic panels. **(HP-I)**

We Support

ASE | **Education Foundation**

Tools and Equipment

Urethane bumper
#80, #180, and #320 grit sandpaper
Appropriate sanding blocks
DA sander
3-inch angle grinder
Aluminum tape
Plastic cleaner
Body filler spreader
Appropriate plastic adhesive cartridge,
    tips, and dispenser
Adhesion promoter (if required by the
    product manufacturer)
Mesh reinforced backing strips
Mixing board or cardboard

Safety Equipment

Safety glasses or goggles
Dust respirator
Work gloves

Introduction

Since the introduction of plastic bumper covers, there have been many changes in the composition of types of plastics used on automobiles. You must first determine the type of plastic you are working with before you can safely and effectively repair it. Some plastics can only be welded, and some may only be repaired using adhesives. For this exercise, we will be using adhesives. There will be two types of repair in this exercise. You will repair a gouge in a bumper for a single-sided repair and a tear in the plastic for a two-sided repair.

Vehicle Description

Year _____ Make _____ Model _____

VIN _____

Procedure

Single-Sided Repair

Task Completed

1. Wash the bumper with soap and water. Clean with a plastic cleaner. ☐

2. Have your instructor make a 6-inch gouge in the bumper using a die grinder with a burr bit or a Dremel could also work. ☐

3. Take your DA sander with #80 grit sandpaper, bevel and featheredge the front side of the tear with #80 grit paper, and sand back all the paint 3–4 inches away from the area of damage. Make sure it is bare plastic before continuing to the next step. ☐

   a. Why should you remove all paint down to the bare plastic? _____
   _____

4. Take your air blower, and blow off the sanded area getting rid of all debris. **Make sure you do not use any cleaner on these areas again.** ☐

   a. Explain what would happen if you were to use cleaner again. _____
   _____
   _____

5. Spray on your adhesion promoter, and follow instructions on the can before applying the repair adhesive. ☐

6. Make sure to equalize two parts of the plastic repair cartridge by loading it into the applicator gun. Squeeze the material out until both sides come out equally together. Hit the release button on the applicator gun to stop the flow. Load a static mixing onto the end of the cartridge, and tighten down the retainer nut if applicable. Then over a mixing board, squeeze an inch of material through the tip before applying onto the surface needing the repair. This ensures that both sides are equally mixed. Now apply the repair adhesive to the surface needing repair and smooth it down with a spreader. Make sure to slightly overfill and allow the material to cure. Do not get any of the material on unsanded areas. ☐

   a. What is the reason for this? _____
   _____

7. Once the adhesive is cured, sand the repair material with #80 grit paper and appropriately sized sanding blocks to remove the roughness. Complete the sanding with #180 grit, and then finish sand with #320 grit until it is smooth and level. ☐

8. If you are unsure if it is level or not, use a paper towel under your hand and feel the area with the towel. It will amplify your sense of feel by allowing you to feel the surface instead of the sand scratches and different surfaces. ☐

9. Once the area is sanded smooth and level, have your instructor check your work. ☐

**Task
Completed**

Begin the Two-Sided Repair Now

The two-sided repair steps are the same for the front of the repair on a single-sided repair as listed earlier. You must reinforce the backside of the tear first, before you repair the front side. This is what gives the repair its strength.

1. Have your instructor use a knife or a cut off wheel to make a 4-inch cut in the bumper. This will simulate a tear. ☐

2. Clean and sand the backside 3–4 inches around the tear with #80 grit sandpaper. ☐

   a. What is the reason for this? _____

   _____

3. Bevel and featheredge the front side of the cut with #80 grit paper. ☐

   a. What is the reason for beveling the front and not the back? _____

   _____

4. Spray on your adhesion promoter, and follow instructions on the can before applying the repair adhesive. ☐

5. Most adhesive repair manufacturers now make a reinforcement cloth to use on the backside of the repair. Carefully cut a piece of the cloth to a similar shape of the tear that covers 1–2 inches around the tear. Use aluminum tape on the front side to maintain alignment. Apply a coat of plastic repair adhesive to the backside of the repair area. Stay within the sanded area. ☐

   a. Why is it important to stay within the sanded area? _____

   _____

6. Now mix and apply the repair adhesive in the same procedure as listed earlier in the single-sided steps. ☐

7. Lay the cloth on top of the repair adhesive. Work the cloth into the flexible repair material with a body filler spreader. If another layer is needed to cover up the cloth, apply as needed before it cures. ☐

   a. Why is it important to completely work the cloth into the repair material?

   _____

8. After the flexible material on the backside has hardened, remove the aluminum tape from the front side. ☐

9. Now repeat the same steps listed earlier in the single-sided repair section for the front side of this repair. ☐

INSTRUCTOR'S COMMENTS _____

_____

_____

Name _____ Date _____

Nitrogen-Welded Repairs—Single Sided and Two Sided

Objective

Upon completion of this activity sheet, the student should be able to safely and accurately repair scratches and tears in urethane bumpers with a nitrogen welder.

ASE Education Foundation Task Correlation

II.A.1   Select and use proper personal safety equipment; take necessary precautions with hazardous operations and materials in accordance with federal, state, and local regulations. **(HP-I)**

II.A.4   Select and use a NIOSH-approved air purifying respirator. Inspect condition and ensure fit and operation. Perform proper maintenance in accordance with OSHA regulation 1910.134 and applicable state and local regulations. **(HP-I)**

II.F.1   Identify the types of plastics; determine repairability. **(HP-I)**

II.F.2   Clean and prepare the surface of plastic parts; identify the types of plastic repair procedures. **(HP-I)**

II.F.3   Repair rigid, semirigid, and flexible plastic panels. **(HP-I)**

We Support

ASE | Education Foundation

Tools and Equipment

Urethane bumper
#80, #180, and #320 grit sandpaper
Appropriate sanding blocks
DA sander
3-inch angle grinder
Aluminum tape
Plastic cleaner
Body filler spreader
Assorted plastic welding rods
Nitrogen plastic welder

Safety Equipment

Safety glasses or goggles
Dust respirator
Work gloves

Introduction

Since the introduction of plastic bumper covers, there have been many changes in the composition of types of plastics used on automobiles. You must first determine the type of plastic you are working with before you can safely and effectively repair it. Some plastics can only be welded, and some may only be repaired using adhesives. For this exercise we will be using adhesives. There will be two types of repair in this exercise. You will repair a gouge in a bumper for a single-sided repair and a tear in the plastic for a two-sided repair.

This is a procedure that you really need to see done before you begin, to help visualize what needs to be done. This will cut down on waste of the welding rods as well. There are a few YouTube videos to watch, all from the Polyvance channel. This is one of the leading companies in nitrogen plastic welding today, and they have tons of videos you can also watch later, especially for some job- and vehicle-specific repairs that are difficult to repair.

Search the Polyvance YouTube channel for the following videos:

- "Hot Air Plastic Welding: Nitrogen vs. Air"
- "Basic Nitrogen Welding Process—TAPAS"
- "Repair a Tear to the Edge of a Bumper Cover with a Nitrogen Plastic Welder"
- "Nitrogen Plastic Welding: Bumper Flange Repair"
- "Bumper Cover Repair—How to Repair Broken Slot Tabs"
- "How to Use Polyvance's 6146 Bumper Pliers Kit"

Vehicle Description

Year _____ Make _____ Model _____

VIN _____

Procedure

Two-Sided Repair

Task Completed

1. Wash the bumper with soap and water. Clean with a plastic cleaner. ☐

2. Have your instructor make a 6-inch tear in the bumper using a die grinder with a cut off wheel, or a knife could also work. ☐

3. Identify the plastic. Find the ID symbol molded into the part, and select the matching welding rod. PP (polypropylene, aka TPO or TEO) is the most common type, but PE (polyethylene) and PC (polycarbonate) are also quite common. ☐

4. Work out any distortions with heat and pressure. Clean the plastic with a plastic cleaner again, and then grind a tapered V-groove into the backside to match the width of the selected welding rod. ☐

5. Take your air blower, and blow off the sanded area getting rid of all debris. **Make sure you do not use any cleaner on these areas again.** ☐

    a. Explain what would happen if you were to use cleaner again. _____

    _____

    _____

6. Use aluminum tape to the front side of the tear to hold the damage in alignment as you weld the backside. ☐

7. Run the backside weld starting 1 inch before the tear and ending 1 inch after the tear. Once the backside weld has cooled, peel the aluminum tape from the front. ☐

    REPAIR NOTE: If the repair you are performing is a tear at the edge of a bumper, you will have to use the T-weld method on the backside of the bumper. Reference the video you watched earlier to perform this task. This will give it the strength it needs to not tear again in the same location.

8. Take your DA sander with #80 grit sandpaper, featheredge the front side of the tear with #80 grit paper, and sand back all the paint 3–4 inches away from the area of damage. Make sure it is bare plastic before continuing to the next step. ☐

    a. Why should you remove all paint down to the bare plastic? _____

    _____

Task Completed

9. V-groove the front side at least halfway into the plastic with a die grinder or belt sander, about the same width as the selected welding rod that you will be using. ☐

10. Now use the nitrogen plastic welder to melt the welding rod and the base material of the plastic panel together, applying a slight downward pressure on the welding rod to the plastic part. If necessary, apply multiple passes with the welding rod to completely fill the V-grooved area. ☐

11. Once the weld has completely cooled down, sand the repair material with #80 grit paper and appropriately sized sanding blocks to remove the roughness. Complete the sanding with #180 grit, and then finish sand with #320 grit until it is smooth and level. ☐

12. If you are unsure if it is level or not, use a paper towel under your hand and feel the area with the towel. It will amplify your sense of feel by allowing you to feel the surface instead of the sand scratches and different surfaces. ☐

13. You may need to apply a skim coat of two-part finishing adhesive as needed to finish the repair and to prepare the surface for primer. **Do not use two-part spot putty or body filler!** It will not properly adhere to the plastic. ☐

14. Once the area is sanded smooth and level, have your instructor check your work. ☐

Single-Sided Repair

1. To perform a single-sided weld is to follow the same steps listed earlier **WITHOUT** the backside repair procedures. Follow the video steps you watched earlier for reference. ☐

Tab Repair

1. To perform a tab repair, reference the videos you watched earlier. Use the same steps as you would for a two-sided weld repair. ☐

2. Use the aluminum tape on the backside of the tab to hold the alignment properly in place and to not allow the melting plastic rod to fall through the hole you are repairing. ☐

   a. Will the plastic welding rod stick to the aluminum tape?

   _____

3. Make sure to allow each side to cool properly before running another weld. ☐

   a. What will happen if you do not allow each side to cool properly? _____

   _____

4. Once both sides have been welded and cooled, it is time to reshape them. ☐

5. Take an angle die grinder to rough shape the tab back to the original shape. ☐

6. Then finish sanding and shaping the tab with a DA sander and #80 and then #180 grit sandpaper. ☐

   a. What would you have to do if you grind or sand too much of the weld off and it is not as strong or as large as it should be shape wise? _____

   _____

7. Redrill the hole with the appropriately sized drill bit. Make sure it is as close to the original shape as possible so that the original mounting hardware will fit properly. ☐

8. Featheredge any remaining sand scratches. ☐

9. Have your instructor check your welds. ☐

INSTRUCTOR'S COMMENTS _____

_____

_____

Name _____ Date _____

Fiberglass Repairs

Objective

Upon completion of this activity sheet, the student should be able to safely and accurately repair fiberglass panels.

ASE Education Foundation Task Correlation

II.A.1	Select and use proper personal safety equipment; take necessary precautions with hazardous operations and materials in accordance with federal, state, and local regulations. **(HP-I)**
II.A.4	Select and use a NIOSH-approved air purifying respirator. Inspect condition and ensure fit and operation. Perform proper maintenance in accordance with OSHA regulation 1910.134 and applicable state and local regulations. **(HP-I)**
II.F.1	Identify the types of plastics; determine repairability. **(HP-I)**
II.F.2	Clean and prepare the surface of plastic parts; identify the types of plastic repair procedures. **(HP-I)**
II.F.3	Repair rigid, semirigid, and flexible plastic panels. **(HP-I)**
II.F.4	Remove or repair damaged areas from rigid exterior composite panels. **(HP-G)**
II.F.5	Replace bonded rigid exterior composite body panels; straighten or align panel supports. **(HP-G)**

We Support

ASE | Education Foundation

Tools and Equipment

Fiberglass panel
Fiberglass cloth, mat, and resin
#80, #180, and #320 grit paper
Appropriate sanding blocks
Body filler spreader
Hammer
Aluminum tape
Cutoff tool
Wax and grease remover

Safety Equipment

Safety glasses or goggles
Dust respirator
Work gloves

Introduction

Late model vehicles have many parts made of plastic, fiberglass, SMC, urethane, and so on. These are strong, durable, and lightweight and do not rust. Because they are very costly to replace, it is important that quality repairs be made. In this job sheet, you will learn how to repair fiberglass panels.

Vehicle Description

Year _____ Make _____ Model _____

VIN _____

Procedure

Scratch

1. Clean the panel with soap and water and then a wax and grease remover. Have your instructor make a 6-inch scratch deep into the plastic. ☐

2. Sand the area around the scratch with #80 grit paper. Mix and apply fiberglass body filler or regular body filler and apply. Overfill the scratch. Allow the material to cure. ☐

   a. What is the purpose of overfilling?_____

   _____

3. Sand the filler to level using #80 grit paper. Featheredge the paint with #180 grit paper, followed by #320 grit. ☐

Break

4. Have your instructor use a hammer to break the fiberglass. Clean the backside of the panel using the same steps as listed earlier, and then sand the backside of the panel with #80 grit paper in an area at least 3 inches to either side of the break. ☐

5. Align the broken area with aluminum tape on the front side. ☐

6. Cut a total of three pieces of fiberglass cloth; one should be the same size as the repair area, and the next two pieces should be slightly larger than the first. ☐

7. Mix the fiberglass resin and hardener according to the product manufacturer's instructions. ☐

8. Apply a layer of adhesive with a paint brush. First put the smallest fiberglass cloth piece over the adhesive. Push it into place with the paint brush or a body filler spreader. Lay down another coat of adhesive and then the second piece of fiberglass cloth that is slightly larger. Apply more adhesive, followed by the last piece of fiberglass. Cover the top with adhesive. Press the adhesive into the fiberglass with the spreader. Allow it to cure. ☐

9. Remove the aluminum tape from the front side. Sand off the paint with the #80 grit paper. Mix up the fiberglass material, and repeat the same steps for the front side. Overfill the low areas. ☐

10. Now sand the material level using #80 grit paper. Featheredge with #180 grit paper, followed by #320 grit. ☐

INSTRUCTOR'S COMMENTS _____

_____

_____

Name _____ Date _____

SMC Repairs

Objective
Upon completion of this activity sheet, the student should be able to safely and accurately repair SMC and fiberglass panels.

ASE Education Foundation Task Correlation

II.A.1	Select and use proper personal safety equipment; take necessary precautions with hazardous operations and materials in accordance with federal, state, and local regulations. **(HP-I)**
II.A.4	Select and use a NIOSH-approved air purifying respirator. Inspect condition and ensure fit and operation. Perform proper maintenance in accordance with OSHA regulation 1910.134 and applicable state and local regulations. **(HP-I)**
II.F.1	Identify the types of plastics; determine repairability. **(HP-I)**
II.F.2	Clean and prepare the surface of plastic parts; identify the types of plastic repair procedures. **(HP-I)**
II.F.3	Repair rigid, semirigid, and flexible plastic panels. **(HP-I)**
II.F.4	Remove or repair damaged areas from rigid exterior composite panels. **(HP-G)**
II.F.5	Replace bonded rigid exterior composite body panels; straighten or align panel supports. **(HP-G)**

We Support

ASE | Education Foundation

Tools and Equipment
SMC panel
Rigid repair material
#80, #180, and #320 grit paper
Appropriate sanding blocks
Body filler spreader
Hammer
Aluminum tape
Cutoff tool
Wax and grease remover

Safety Equipment
Safety glasses or goggles
Dust respirator
Work gloves

Introduction
Newer model vehicles have many parts made of plastic, fiberglass, SMC, urethane, and so on. These are strong, durable, and lightweight and do not rust. Because they are very costly to replace, it is important that quality repairs be made. In this job sheet, you will learn how to repair SMC panels.

Vehicle Description

Year _____ Make _____ Model _____

VIN _____

NOTE: For an SMC tear, you will need to follow the same steps as the plastic adhesive repair shown in Job Sheet 13.1, except for replacing the plastic adhesive material for the SMC repair adhesive.

Procedure

SMC Splice

Task Completed

1. Cut a 2-inch × 4-inch rectangle out of SMC panel with a cutoff tool. Cut out two 1-inch × 5-inch backers from a scrap piece of SMC. ☐

2. Clean and sand the backside of the hole and the backers with #80 grit. Mix rigid repair material adhesive. Apply to the long sides of the rectangle. Place the backers in the adhesive. Overlap the hole by ½ inch. Allow proper cure time. ☐

    a. What is the purpose of the backers?

    _____

3. Clean and sand both sides of the cut-out rectangle. Mix up the rigid repair material adhesive and spread a coat on the overlapped portion of the backers. Place the cut-out rectangle on the backers. Allow this to cure. ☐

4. Clean, sand with #80 grit, and bevel the area around the hole. ☐

    a. What is the reason for beveling the area? _____

    _____

    b. What problems might result if you do not bevel the area? _____

    _____

5. Mix the repair material, and overfill the gaps between the hole and the cut-out piece. ☐

    a. What is the reason for overfilling the area? _____

    _____

6. Allow cure time. Sand the filler with #80 grit. Featheredge with #180, and then finish with #320 grit. ☐

INSTRUCTOR'S COMMENTS _____

_____

_____

Name _____ Date _____

Shrinking and Shaping Plastic

Objective

Upon completion of this activity sheet, the student should be able to safely and accurately shrink and reshape damaged urethane bumpers.

ASE Educational Foundation Task Correlation

II.A.1	Select and use proper personal safety equipment; take necessary precautions with hazardous operations and materials in accordance with federal, state, and local regulations. **(HP-I)**
II.A.4	Select and use a NIOSH-approved air purifying respirator. Inspect condition and ensure fit and operation. Perform proper maintenance in accordance with OSHA regulation 1910.134 and applicable state and local regulations. **(HP-I)**
II.F.1	Identify the types of plastics; determine repairability. **(HP-I)**
II.F.3	Repair rigid, semirigid, and flexible plastic panels. **(HP-I)**

We Support

Education Foundation

Tools and Equipment

Urethane bumper
#80, #180, and #320 grit sandpaper
Heat gun
Paint sticks
Screwdrivers

Safety Equipment

Safety glasses or goggles
Dust respirator
Work gloves

Introduction

Since the introduction of plastic bumper covers, there have been many changes in the composition of types of plastics used on automobiles. You must first determine the type of plastic you are working with before you can safely and effectively repair it. Some plastics can only be heated and shaped one time, and then they are permanently locked in place. Others can be reheated and reshaped over and over until the desired shape has been achieved. This job sheet will help you achieve the desired result with practice.

Vehicle Description

Year _____ Make _____ Model _____

VIN _____

Procedure

Shrink

1. First, you must identify the type of plastic you need to repair. Then determine if it can be shrunk. ☐

2. A stretched bumper cover can be recognized by a bulging or swollen area. Locate a stretched area on a bumper. ☐

   a. Describe the damage of your bumper; does it bulge in or out? _____

   _____

   _____

   b. Would that make it concave or convex? _____

3. Using a heat gun set on high, warm up an area 3–4 inches larger than the swollen area. The plastic will heat rapidly, so be careful. ☐

   a. What can happen if too much heat is applied to one area? _____

   _____

   _____

4. When the plastic has heated up, you may have to help shape it back into place. You may use any tool necessary such as the rounded end of a screwdriver or a paint stick for flat areas. Many other tools can be used depending on the shape of the damaged area. Use your own discretion, but you must remember to not push hard on any tool initially so that you do not push through the plastic.

5. Once shaped, rapidly cool it with a wet rag, ice, or compressed air. ☐

   a. What happens to the panel when it is cooled? _____

   _____

   _____

6. Repeat this process as necessary with the remaining swollen areas.

Shape

7. First, you must identify the type of plastic you need to repair. Then determine if it can be reshaped. ☐

8. Collision-damaged bumpers are often bent out of shape. Heat can be used to reshape them. Obtain a collision-damaged urethane bumper. ☐

9. Examine the damage. Decide where the plastic is damaged and where it needs to be moved. ☐

   a. What is the damage on your panel? _____

   _____

   b. How do you decide when a bumper can be repaired or when it needs to be replaced?

   _____

10. Warm the first area to be moved, using a heat gun set on high. When the plastic
    is hot to the touch, it can be moved into its proper location with a paint stick.

    a. What precautions must be taken when checking the temperature of the
        bumper?

    _____

    _____

11. When the damaged area is in the right location, rapidly cool the area with a wet
    rag or ice. Repeat the process on other damaged areas.

    a. What safety precautions should be taken when using an electric heat gun
        near wet rags?

    _____

    _____

INSTRUCTOR'S COMMENTS _____

_____

_____

Review Questions

Name _____ Date _____ Instructor Review _____

1. Some plastic repair materials require a(n) _____ to make the adhesive stick.

2. You need to know the ISO code before determining what repair procedure to use.
   A. True
   B. False

3. You must clean the bumper thoroughly before sanding and before applying the adhesion promoter.
   A. True
   B. False

4. When repairing urethane, Technician A uses a flexible repair material. Technician B uses a rigid repair material. Who is correct?
   A. Technician A
   B. Technician B
   C. Both Technician A and Technician B
   D. Neither Technician A nor Technician B

5. When repairing SMC, Technician A uses body filler. Technician B uses rigid repair material. Who is correct?
   A. Technician A
   B. Technician B
   C. Both Technician A and Technician B
   D. Neither Technician A nor Technician B

6. The argon gas from a nitrogen welder acts as a shielding gas around the weld site and allows for a noncontaminated weld.
   A. True
   B. False

7. Tab repair can be done by using adhesives but not by a nitrogen welder.
   A. True
   B. False

8. When splicing SMC, Technician A uses a backing strip. Technician B leaves a ¼-inch gap between the panels. Who is correct?
   A. Technician A
   B. Technician B
   C. Both Technician A and Technician B
   D. Neither Technician A nor Technician B

9. Technician A states that a two-sided weld is the strongest. Technician B believes that when making a two-sided weld, one must V-groove both sides. Who is correct?
   A. Technician A
   B. Technician B
   C. Both Technician A and Technician B
   D. Neither Technician A nor Technician B

10. Technician A says that a T-weld is recommended for a tear on the edge of a bumper cover. Technician B says that it is recommended for repairing tabs. Who is correct?
   A. Technician A
   B. Technician B
   C. Both Technician A and Technician B
   D. Neither Technician A nor Technician B

CHAPTER 14

Passenger Compartment Service

Name _____ Date _____ Instructor Review _____

Troubleshooting Noise Leaks

In the chart that follows, write in the corrections to the problems listed. Do not use your textbook.

Sources	Cause	Corrections
1. Weatherstrip	Imperfect adhesion to contact surface and improper contact of lip due to separation, breakage, crush, and hardening .	
2. Door sash and related parts	1. Improper weatherstrip contact due to a bent door sash 2. Gap caused by improperly installed corner piece 3. Gap caused by badly finished corner sash 4. Separation and breakage of the rubber on the door glass run	
3. Door assembly	Improper weatherstrip contact due to improperly fitting door	
4. Door glass	Gap caused due to ill-fitting door glass	
5. Body	Improper body finishing on contact surface for door weatherstrip (uneven panel joint, sealer installed improperly, and spot welding splash)	
6. Drip molding	Rise and separation of molding	
7. Front pillar	Rise and separation of molding	
8. Waist molding	Door glass gap due to rise of molding and deformation of rubber seal	

**SHOP ASSIGNMENT
14-2**

Name _____ Date _____ Instructor Review _____

Identification of Instrument Panel Parts

Without using your textbook, write in the names of the numbered parts. When complete, show your instructor. If you have missed any, use your textbook to identify them and also give their functions.

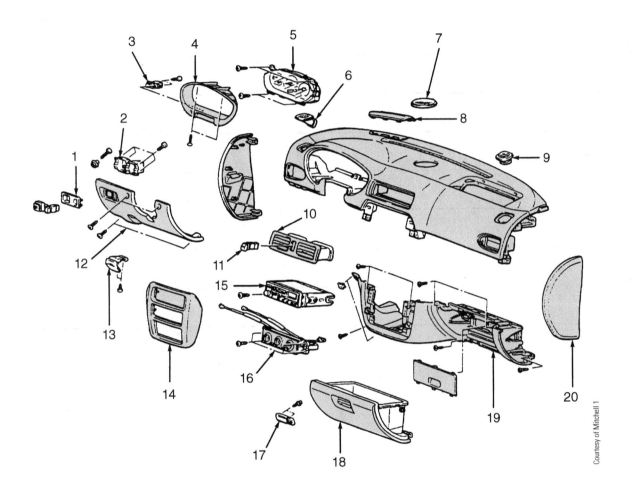

Courtesy of Mitchell 1

1. _____

2. _____

3. _____

4. _____

5. _____

6. _____

7. _____

8. _____

9. _____

10. _____

11. _____

12. _____

13. _____

14. _____

15. _____

16. _____

17. _____

18. _____

19. _____

20. _____

Name _____ Date _____ Instructor Review _____

Seat Part Identification

Using the following figure, name the parts.

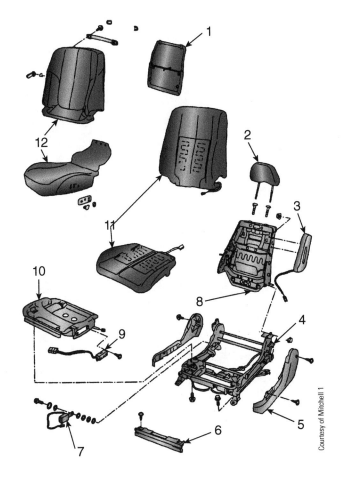

Courtesy of Mitchell 1

1. _____

2. _____

3. _____

4. _____

5. _____

6. _____

7. _____

8. _____

9. _____

10. _____

11. _____

12. _____

Name _____ Date _____

Remove and Install Front and Rear Seat

Objective

Upon completion of this activity sheet, the student should be able to safely inspect, remove, and reinstall the seats in almost any vehicle.

ASE Education Foundation Task Correlation

II.B.3 Inspect, remove, label, store, and reinstall interior trim and components. **(HP-I)**

II.B.4 Inspect, remove, label, store, and reinstall body panels and components that may interfere with or be damaged during repair. **(HP-I)**

II.B.5 Inspect, remove, protect, label, store, and reinstall vehicle mechanical and electrical components that may interfere with or be damaged during repair. **(HP-G)**

II.B.6 Protect panels, glass, interior parts, and other vehicles adjacent to the repair area. **(HP-I)**

We Support

ASE | **Education Foundation**

Tools and Equipment

Vehicle
Flashlight
Appropriate hand tools

Safety Equipment

Work gloves
Safety glasses or goggles

Introduction

Seats must provide comfort as well as safety for the passenger day after day as well as on long trips. Some are simple bench seats; some are electric, providing many seat settings for passenger comfort; and some are heated and cooled. Seats are made of leather, cloth, and vinyl. When removing seats, it is important that no damage occurs to the seat or the vehicle's paint while pulling the seats out.

If you have any difficulty or doubt about any aspect of this activity, be sure to ask your instructor for help or to answer your questions.

Vehicle Description

Year _____ Make _____ Model _____

VIN _____

Procedure

Task Completed

1. With your flashlight check under the seats to make sure nothing will be damaged. Start with the driver's seat. Move it either all the way forward or all the way back to expose the channel bolts. Sometimes these bolts are covered. Use a ⅜-inch ratchet and appropriate socket to loosen the bolts. Sometimes it is necessary to use a ⅜-inch air ratchet or battery impact for faster removal. Most newer vehicles have plastic trim covers to hide the appearance of the bolts/nuts. Locate and remove these covers carefully without breaking them to remove the bolts/nuts. Some vehicles have bolts sticking up through the bottom of the floor with the seats held by nuts. Some have bolts that secure the seat to the floor. If the bolts are rusted, do not force them. Use some penetrating oil that is body shop safe. □

DO NOT use WD-40 as this will cause airborne contaminants and create fish eye in the paint booth area.

a. Are there nuts or bolts securing the seat? _____

2. When the nuts/bolts are loose, move the seat in the opposite direction to get to the other nuts or bolts. These should be the same size. Once off, this should release the seat from the floor. **DO NOT** try to pull out the seat yet. Check to see if the seat belts are run through the seat or not. If so, they must be pulled through the seat first. These are usually a very tight fit and must not be forced out. If they are separate from the seat, you may remove. ☐

a. Is any part of the seat belt bolted onto the side of the seat? _____

3. On power seats, there will be a plug under the seat. Locate the plug and disconnect it. Sometimes there is an air bag harness connector attached under a seat for circumstances when there is a side air bag inside the seat. It will be the yellow connector. Some models have an orange protective clip you must slide out of the air bag connector; then you can disconnect the harness connector. ☐

a. Was there an air bag harness or only a power seat harness? _____

4. Once everything is disconnected, have a partner help you remove the seat from the vehicle. Be very careful when removing the seat so you do not scratch the paint on the vehicle or damage the seat. When the seat is out, set it out of the way. If you have to leave it out for a while, you might want to consider covering it with some plastic to keep dust and debris from getting on it. ☐

5. Repeat the same steps for the passenger side of the vehicle. With both seats out, check the condition of the carpet. ☐

a. Describe any previous damage to the seats or damage that may have occurred to the seats or door openings. _____

_____

This Section Is for Back Seats in Most Cars: If you are working on a truck or an SUV, please go to the next section.

6. With the front seats out, you should have good access to get the back seat out. Most back seat bottoms are held on with hooks that latch onto metal clips in the floor. While kneeling on either side, look where the floor pan meets seat and you should see a pull tab or a plastic clip to pull. Sometimes you may have to **SLIGHTLY** lift up on the front of the bottom seat to visibly see this. Pulling too hard may result in damaging the seat bottom. This should release the seat. Seat belts must be pulled through the back seat after the seat is removed. ☐

7. With the bottom of the back seat out, examine the top side of the back seat. ☐

a. What is holding the top of the seatback to the package shelf frame? _____

**Task
Completed**

8. Once it is loose, lift up on the seat to remove it and then place it on a clean table. Check the condition of the seat. ☐

a. Is there any damage to the seat? If so, describe the damage.

_____

_____

This Section Is for Back Seats in Most SUVs and Trucks

Most SUV and truck back seats are very similar to the front seats as far as the way they are bolted in. They are rarely power seats but can still have wiring harnesses run to them for seat belt and/or air bag sensors.

9. Lift up the bottom seat and raise it forward. Underneath the bottom of the seat will be the bolts that hold the seat frame in. Sometimes there are plastic covers or trim pieces that hide the bolts. Being careful not to break anything, remove these covers and use the appropriate size sockets and ratchet to remove these. Once removed, disconnect any wiring harnesses that are connected to the seats. Then remove the back seats with the help of a partner so that no damage occurs to the seats or the painted panels around the door openings. ☐

10. When all the seats are out, it is a good time to vacuum thoroughly and even shampoo (if necessary) the carpet for the customer. It does not take much time or money to do this, and it will make for great customer relations.

11. After the seats and interior are clean, reinstall the seats in the reverse order that you removed them. Install the top half of the back seat. Before installing the seat bolts, make sure the seat belts are secured in their proper position. Hand-start and hand-tighten the bolts first. Once the bolts are hand-tightened, finish tightening with the proper ratchet and socket. ☐

a. What is the reason for only hand-tightening at first? _____

_____

12. Place the bottom half of the seat into position in the car. Lift up the front of the seat while pushing it back. Insert the seat belts through their proper holes in the seat. Once this is done, lock the seat the same way it was released. ☐

a. Did the seat secure all the way? _____

b. Do all the seat belts function properly? _____

13. After the back seat is secured, carefully install the passenger seat. Feed the seat belt through the seat. Make sure the bolt holes are aligned, and then hand-start and hand-tighten only the two back bolts. When this is complete, move the seat all the way back and insert the front bolts. After all the bolts are aligned and tightened, use the proper ratchet and socket to tighten. ☐

a. Were any of the bolts stripped? _____

b. If so, how will you make this repair? _____

_____

14. Finally, install the driver's seat. Position the seat, if applicable feed the seat belt, and connect any wires before nuts are tightened. Use the same procedure to reinstall the passenger's seat. ☐

**Task
Completed**

15. After all the seats are installed, sit in each one while checking its operation.   ☐

    a. Does each seat move freely? _____

    b. If there is binding somewhere, how will you make this repair? _____

    _____

    c. Do all the seat belts function properly? _____

16. If not, make the necessary repairs so that the seats and seat belts function   ☐
    properly. When all parts are functioning properly, install all caps and other
    interior trim that cover seat parts.

INSTRUCTOR'S COMMENTS _____

_____

_____

Name _____ Date _____

Remove and Install Headliner

NOTE: YOU MUST HAVE COMPLETED Job Sheet 14.1
BEFORE STARTING THIS JOB SHEET!

Objective

Upon completion of this activity sheet, the student should be able to safely inspect, remove, and reinstall the headliner in almost any vehicle.

ASE Education Foundation Task Correlation

II.B.3	Inspect, remove, label, store, and reinstall interior trim and components. **(HP-I)**
II.B.4	Inspect, remove, label, store, and reinstall body panels and components that may interfere with or be damaged during repair. **(HP-I)**
II.B.5	Inspect, remove, protect, label, store, and reinstall vehicle mechanical and electrical components that may interfere with or be damaged during repair. **(HP-G)**
II.B.6	Protect panels, glass, interior parts, and other vehicles adjacent to the repair area. **(HP-I)**

We Support

ASE | Education Foundation

Tools and Equipment

Vehicle
Flashlight
Appropriate hand tools

Safety Equipment

Work gloves
Safety glasses or goggles

Introduction

Headliners are the large fabric-covered assemblies that are installed into automobiles to hide the underside of roof panels and to add extra sound deadening to the interior of the vehicle. To remove these, carefully follow the steps listed next. Make sure to take care as you remove all components not to tear or break anything. These assemblies are very expensive and sometimes require the front or back glass of the vehicle to be removed before you can remove/install the headliners.

Be sure to ask your instructor if you have any difficulty, doubt, or questions with any aspect of this activity.

Vehicle Description

Year _____ Make _____ Model _____

VIN _____

**Task
Completed**

Procedure

1. All vehicles are different and require different procedures for removing the headliner. At times, seats can get in the way of removing the trim from the pillars. If so, you should have completed job sheet 14.1 before beginning. Refer to those instructions on how to remove the seats as needed.   ☐

Task Completed

2. After the seats are removed, carefully examine every panel that is attached to the headliner. These must all be removed.   ☐

3. Most visors are attached with Phillips or Torx screws. Once these are removed, make sure to properly label these screws in their own separate container or sealable bag.   ☐

4. If there is a dome light or reading lights, sometimes they will have to be removed. Refer to the OEM service info for specific make and model instructions, or a quick YouTube search can be helpful. Carefully remove them by taking out the plastic lens to reveal the attaching components.   ☐

   a. Did the dome light have to be removed on this model, or was it made into the headliner? _____

5. If the dome light must be removed, loosen the dome light and gently pull it down until the wires are exposed. There should be a connector to disconnect the light.   ☐

6. If there are reading lights, remove them using the same method you used to remove the dome light.   ☐

   a. Did the reading lights have to be removed on this model, or was it made into the headliner? _____

   _____

7. Look to see if there are any garment hangers or grab handles in the headliner. These must also be removed. If the vehicle you are working on has these parts, you may have to take a pick or a nonmarring tool to remove the screw covers to expose the retaining screws that hold them in. Make sure to label these separately as well so they do not become confused with other trim screws.   ☐

8. Usually the pillar trim overlaps the trim around the windshield and doors. Remove the screws, but do not pull on the trim hard. There sometimes can be clips holding the trim on as well. Once removed, make sure to label these separately as well so they do not become confused with other trim screws.   ☐

9. At this point, it is a good idea to have a partner in the vehicle to hold up the liner. Remember, this panel is fragile and can crack or wrinkle easily. With all the pillar trim off, examine and remove the moldings that are holding the liner in place.   ☐

10. The headliner should be completely loose and ready to remove from the vehicle. With the help of your partner, **CAREFULLY** work the panel through the largest opening.   ☐

    a. Was there any damage that occurred in the removal process? _____

    b. Did you properly label all trim screws after removal? _____

Task Completed

Reinstallation

11. With your partner, slowly work the headliner into the vehicle. Some wiring connectors must be connected before the headliner is set into place. When it is in place, have your partner hold it while you attach the trim molding at the windshield and then at the back window. There should be clips to lock the trim in place before the screws are installed. Make sure all wires are threaded through their proper openings.   ☐

    a. Why are the front and rear trim panels installed first? _____

    _____

    _____

12. Do not tighten any screws at this point. Start reattaching the pillar trim pads in reverse order.   ☐

    a. Explain why it is important to only hand-start all the securing screws at first.

    _____

    _____

    _____

13. Make sure all the trim panels are in place and not cracked. Once finished, tighten all the screws.   ☐

14. Reattach all the hangers and handles and then the dome and reading lights.   ☐

    a. Do all the lights work properly? _____

    b. If not, how will you correct the problem? _____

15. Reinstall the seats, if necessary, and any other trim pieces that were removed.   ☐

INSTRUCTOR'S COMMENTS _____

_____

_____

Review Questions

Name _____ Date _____ Instructor Review _____

1. The _____ is a cloth or vinyl cover for the inside of the roof panel.

2. The dash panel assembly is also known as the _____.

3. _____ normally cause a whistling in the passenger compartment when the vehicle is driven.

4. Sill plates are also called scuff plates.
   A. True
   B. False

5. The majority of leaks occur at the top of the windshield or top and bottom of the rear window.
   A. True
   B. False

6. A headrest will help avoid whiplash in an accident.
   A. True
   B. False

7. Technician A says that any good bolt can secure the seat track to the floor. Technician B says that case-hardened bolts must be used. Who is correct?
   A. Technician A
   B. Technician B
   C. Both Technician A and Technician B
   D. Neither Technician A nor Technician B

8. Technician A says that the dash assembly is also called the dashboard. Technician B says that the assembly is also called the instrument panel. Who is correct?
   A. Technician A
   B. Technician B
   C. Both Technician A and Technician B
   D. Neither Technician A nor Technician B

9. Technician A says to use a water hose with a spray nozzle to check for water leaks. Technician B says to use a water hose without a spray nozzle. Who is correct?
   A. Technician A
   B. Technician B
   C. Both Technician A and Technician B
   D. Neither Technician A nor Technician B

10. Technician A says that most rattle repairs involve the readjustment of parts. Technician B says that replacement of parts will correct the problem. Who is correct?
   A. Technician A
   B. Technician B
   C. Both Technician A and Technician B
   D. Neither Technician A nor Technician B

CHAPTER 15

Welding Equipment Technology

Name _____ Date _____

MIG Welding

Objective

Upon completion of this activity sheet, the student should be able to safely use a MIG welder. The student should also be able to weld flat, vertical, and overhead welds.

ASE Education Foundation Task Correlation

II.C.16   Weld damaged or torn steel body panels; repair broken welds. **(HP-G)**

VI.A.1   Select and use proper personal safety equipment; take necessary precautions with hazardous operations and materials in accordance with federal, state, and local regulations. **(HP-I)**

VI.A.2   Locate procedures and precautions that may apply to the vehicle being repaired. **(HP-I)**

VI.A.3   Identify vehicle system hazard types (supplemental restraint system [SRS], hybrid/electric/alternative fuel vehicles), locations, and recommended procedures before inspecting or replacing components. **(HP-I)**

VI.A.4   Select and use a NIOSH-approved air purifying respirator. Inspect condition and ensure fit and operation. Perform proper maintenance in accordance with OSHA regulation 1910.134 and applicable state and local regulations. **(HP-I)**

VI.B.2   Determine the correct GMAW welder type, electrode/wire type, diameter, and gas to be used in a specific welding situation. **(HP-I)**

VI.B.3   Set up, attach work clamp (ground), and adjust the GMAW welder to "tune" for proper electrode stickout, voltage, polarity, flow rate, and wire-feed speed required for the substrate being welded. **(HP-I)**

VI.B.4   Store, handle, and install high-pressure gas cylinders; test for leaks. **(HP-I)**

VI.B.5   Determine the proper angle of the gun to the joint and direction of gun travel for the type of weld being made. **(HP-G)**

VI.B.6   Protect adjacent panels, glass, vehicle interior, and so on from welding and cutting operations. **(HP-I)**

VI.B.9 Clean and prepare the metal to be welded, assure good metal fit-up, apply weld-through primer if necessary, and clamp or tack as required. **(HP-I)**

VI.B.10 Determine the joint type (butt weld with backing, lap, etc.) for weld being made. **(HP-I)**

VI.B.11 Determine the type of weld (continuous, stitch weld, plug, etc.) for each specific welding operation. **(HP-I)**

VI.B.12 Perform the following welds: plug, butt weld with and without backing, fillet, and so on in the flat, horizontal, vertical, and overhead positions. **(HP-I)**

VI.B.13 Perform visual evaluation and destructive test on each weld type. **(HP-G)**

VI.B.14 Identify the causes of various welding defects; make necessary adjustments. **(HP-I)**

VI.B.15 Identify cause of contact tip burnback and failure of wire to feed; make necessary adjustments. **(HP-I)**

We Support

ASE | **Education Foundation**

Tools and Equipment
Weld coupons
Weld stand
MIG welder
16-gauge thick metal
Weld clamps
Tape measure

Safety Equipment
Safety glasses or goggles
Welding helmet (auto darkening preferred)
Welding gloves
Welding sleeves or welding jacket
Welding respirator
High-top safety shoes

Introduction
Along with many advances in the automotive field, many developments have also been made in the welding industry. In the collision repair industry, most structural welding must be done with a MIG welder. There are variations such as MIG welding high-strength steel with silicon bronze but is still a MIG welder. MIG welding provides stronger welds with less heat distortion than oxyacetylene welding.

Welder Description
Brand _____ Model _____

Procedure

Task Completed

1. Your instructor will demonstrate how to change the spool and set tension of the wire feed. ☐

    a. Summarize the steps here. _____

    _____

    _____

    b. What is the wire size? _____

    c. What is the wire type? _____

    d. What metal is the type of wire recommended for?

    _____

NOTE: If the welder is out of gas, turn the gas knob all the way off. Loosen the nut at the top of the gas tank, and remove the tank. Replace the tank with a new filled tank. Replace the thread tape on the threads before inserting the nut back onto the new tank. Tighten the nut down tight until snug. Turn the gas knob back on, and check for any leaks.

    e. How do you know if the tank is empty or not? _____

    _____

f.   How can you tell if there are any leaks after replacing the tank? _____

_____

g.   What corrections will you make to fix the leaks? _____

_____

2.   Your instructor will demonstrate the proper way to hold the gun and the proper speed to   ☐
move the gun. Select the metal coupons and practice proper gun speed in pull movements.

a.   What gauge is the metal you are working on? _____

b.   How did you determine the gauge of the metal? _____

_____

c.   How do you determine whether the travel speed is correct? _____

_____

d.   What does it mean to pull? _____

_____

e.   Examine your weld. Does it have a consistent burn-through? _____

_____

3.   Fill out the chart on weld problems.

Problem	How to Correct
Excessive melt-through	
Incomplete penetration	
Warpage	
Porosity	
Incomplete fusion	
Weld metal cracks	

4.   Explain how to perform a destructive test on a weld. _____

_____

_____

Weld a Bead

5.   Cut a 12-inch × 12-inch piece of 16-gauge sheet metal. Lay it flat on a weld table. Attach   ☐
a ground clamp to the piece. Set the gas flow, voltage, and wire feed speed.

a.   How are these settings done? _____

_____

6.   Put on your jacket, helmet, and gloves. Trigger the gun. With proper stickout, move the   ☐
gun at the correct speed. Adjust the wire speed until a steady hiss is heard and a good
bead profile is visible. The welder is now tuned.

7.   Tightly clamp a piece of 12-inch × 6-inch sheet metal to the original piece.   ☐

a.   What can happen if the pieces are not tightly clamped? _____

_____

8.   Hold the gun at a 70-degree angle. Run a bead along the edge to form a lap joint.   ☐

a.   Examine the weld.

9.   Perform a test on this weld. If it does not pass, continue making this weld until it is correct.   ☐

Different Thicknesses of Welds

10. To weld different thicknesses, determine the thickness of the metal you are going to weld. Then, adjust the settings on the welder and tune until you can consistently run a smooth bead. The welder will tell you what settings it should be set at for each thickness of metal. This level of skill takes considerable practice. ☐

Flat Welds
Procedure
Lap Weld

11. Overlap two weld coupons to form a lap joint. Clamp them together tightly. Adjust and tune the welder. ☐

   a.   Why is it important to clamp the pieces together? _____
       _____

   b.   How is a MIG welder adjusted and tuned? _____
       _____

12. Follow the sequence for skip welding given in your workbook. Make a ¾-inch bead at the center of the overlap. Allow the weld to cool. ☐

   a.   How long does it take the metal to cool? _____

13. Weld another ¾-inch bead on the left edge of the plate. Then go back and continue to weld ¾ inch on the right edge of the plate. Continue this sequence until the entire length is welded. Test the weld. If it is not acceptable, practice until the welds will hold. ☐

   a.   How is a weld tested? _____
       _____

Plug Weld

14. Drill ten ⁵⁄₁₆-inch holes in a weld coupon, or punch 10 holes using a ⁵⁄₁₆-inch hole punch. Measure the distance between the first hole and the other nine. ☐

15. Overlap this coupon onto another coupon with no holes in it. Clamp the two coupons together. Adjust and tune the welder. To make this weld, hold the gun at a 90-degree angle and start at the edge of the hole. Weld the edge of the top panel to the bottom panel. Evaluate the appearance of and test the weld. The weld should completely fill in the diameter of the hole and be smooth with the coupon surface. Very little of the weld should be sticking above the surface. If too much is sticking above (if it looks like the shape of a mountain or a piece of chewing gum placed on the surface), repeat steps 4–5 until you can consistently make smooth, minimal height plug welds. ☐

Butt Weld

16. Place a coupon on the weld table. Place two coupons edge to edge with a ¹⁄₁₆-inch gap between them. ☐

   a.   Why is it important to have a gap in the pieces? _____
       _____

17. Clamp the coupons together. Place a tack weld at the center and on each edge. Then use the skip weld technique. The goal here is to have a mostly flush weld from end to end that is consistent in width of the weld, depth penetration, and uniform in shape. Examine and test the weld. ☐

   a.   Why is it important to skip weld on this piece? _____
       _____

18. Perform a test on this weld. If it does not pass, continue making this weld until it is correct. ☐

Butt Joint with Backing

19. Arrange three coupons, one underneath the other two. The butt joint should have a
    1/16-inch gap. Clamp this into a weld stand. Adjust and tune the welder. Tack weld at
    the center and at each end.   ☐

    a.   What is the reason for the 1/16-inch gap? _____

    _____

    b.   Why is the piece tack welded instead of using a continuous weld? _____

    _____

20. Skip weld, working out from the center, always pulling the gun down.   ☐

    a.   Why is it important to work out from the center? _____

    _____

    b.   Why is it important to pull the gun instead of pushing it? _____

    _____

Vertical Welds
Procedure
Lap Weld

21. Clamp two coupons to a welding stand so that the joint is vertical. Adjust and tune the   ☐
    welder. Skip weld the joint, always moving the gun downward from top to bottom.
    Examine and test the weld.

22. Perform a test on this weld. If it does not pass, continue making this weld until it is correct.   ☐

Plug Weld

23. Drill ten 5/16-inch holes in a weld coupon, or punch 10 holes using a 5/16-inch hole punch.   ☐
    Overlap onto another coupon. Clamp the coupons in a welding stand in a vertical
    position. Adjust and tune the welder. Plug weld the holes. Examine and test the weld.

24. Perform a test on this weld. If it does not pass, continue making this weld until it is correct.   ☐

Butt Weld

25. Place a coupon on the weld table. Place two coupons edge to edge with a 1/16-inch gap   ☐
    between them. Clamp the coupons in a welding stand in a vertical position.

26. Perform a test on this weld. If it does not pass, continue making this weld until it is correct.   ☐

Butt Joint with Backing

27. Arrange three coupons, one underneath the other two. The butt joint should have a   ☐
    1/16-inch gap. Clamp this into a weld stand in a vertical position. Adjust and tune the
    welder. Tack weld at the center and at each end.

28. Perform a test on this weld. If it does not pass, continue making this weld until it is correct.   ☐

Overhead Welds
Procedure

29. Clamp two coupons in the overhead position on the welding stand. Overhead welds □
    require lower voltage and lower wire speeds. This produces shorter arc and adds control
    of the weld puddle.

   a.   What machine settings are adjusted to help control the weld puddle in overhead
        welding?_____

   b.   Are machine settings set slightly higher or lower in overhead welding? _____

   c.   Explain why. _____

        _____

        _____

   d.   What is the skip technique? _____

        _____

   e.   Why is the skip technique used? _____

        _____

Lap Weld

30. Overlap two weld coupons to form a lap joint. Clamp them together tightly in the □
    overhead position. Adjust and tune the welder.

31. Perform a test on this weld. If it does not pass, continue making this weld until it is correct. □

Plug Weld

31. Drill 5/16-inch plug holes 1/2-inch apart along the length of the coupon and overlap it on □
    another. Clamp into the weld stand in the overhead position. Adjust and tune the welder.
    Weld the holes. Examine and test the weld.

32. Perform a test on this weld. If it does not pass, continue making this weld until it is correct. □

Butt Weld

33. Place a coupon on the weld table. Place two coupons edge to edge with a 1/16-inch gap □
    between them. Clamp the coupons in a welding stand in the overhead position.

34. Perform a test on this weld. If it does not pass, continue making this weld until it is correct. □

Butt Joint with Backing

35. Position three coupons and clamp in the overhead position with a 1/16-inch gap between □
    them. Tack weld at the center and on either edge.

36. Perform a test on this weld. If it does not pass, continue making this weld until it is correct. □

INSTRUCTOR'S COMMENTS _____

_____

_____

Name _____ Date _____

Oxyacetylene Torch

Upon completion of this activity sheet, the student should be able to safely heat and cut with an oxyacetylene torch.

ASE Education Foundation Task Correlation

VI.B.1   Identify the considerations for cutting, removing, and welding various types of steel, aluminum, and other metals. **(HP-G)**

VI.B.4   Store, handle, and install high-pressure gas cylinders; test for leaks. **(HP-I)**

VI.B.7   Identify hazards; foam coatings and flammable materials prior to welding/cutting procedures. **(HP-G)**

VI.B.8   Protect computers and other electronics/wires during welding procedures. **(HP-I)**

We Support

Education Foundation

Tools and Equipment

Oxyacetylene cutting torch
Rusted on bumper bolts
Spark lighter
Hammer
Appropriate socket and ratchet
Vehicle

Safety Equipment

Safety glasses or goggles
Welding tinted face shield with #5 lens
Welding jacket
High-top safety shoes
Welding gloves
Fire extinguisher

Introduction

Before the introduction of MIG welding, oxyacetylene brazing and welding were the accepted norm in collision centers. Because brazing created much less heat than welding, it was used to secure body parts. Oxyacetylene welding was primarily used for structural parts such as frames. Any cutting of heavy parts was done with a cutting torch.

Procedure (*with instructor present*)

Task Completed

1. Examine the area to be heated. There should not be any electrical wires, gas lines, undercoat, or flammable materials in the way. There should also be room for you to be out from under the heated area. Do not heat gas or fluid-filled isolators.   ☐

    a. What is the red hose for? _____

    _____

    b. What is the green hose for? _____

    _____

2. Your instructor will demonstrate how to turn on the gas valves and light the torch. Set the gauges at 8–25 psi for oxygen and 3–8 psi for acetylene. Open the acetylene valve ⅛ of a turn and use a spark lighter to ignite the torch. Slowly open the oxygen valve.   ☐

    a. What happens if the psi is too high? _____

    _____

    b. What happens to the flame if the psi is too low? _____

    _____

Task Completed

c.   What can happen if a cigarette lighter is used to light a torch? _____

_____

d.   Why is the acetylene only opened 1/8 of a turn? _____

_____

e.   What can happen if the valves are opened too quickly? _____

_____

3.   Adjust the flame to a neutral flame.   ☐

a.   What does a neutral flame look like? _____

4.   Crawl underneath the vehicle.   ☐

a.   Describe precautions necessary to work under a vehicle. _____

_____

5.   Have a partner to hand you a torch. Keep the torch pointed away from you and everything   ☐
else! Keep the torch about 1 inch from the nut. Heat the nut until it grows red. Turn off
the torch and set it out of the way.

a.   To turn off the torch, which gas valve is turned off first? Why? _____

_____

6.   Put the appropriate-sized socket and ratchet on the nut. Try to turn it.   ☐

7.   If the nut does not turn, relight the torch and heat the nut again. Hold the tip of the preheat   ☐
flame approximately 1/8 inch from the bolt, and repeat steps 5–6 until the nut breaks loose.
Be sure to have your partner stand nearby with a fire extinguisher.

a.   What will happen if you pull the oxygen trigger? _____

_____

8.   Turn the torch off and set it aside. Tap the bolt off with a hammer. After the bolt is off,   ☐
check the bolt hole in the bumper.

a.   Is there any damage? If so, how will it be repaired? _____

_____

INSTRUCTOR'S COMMENTS _____

_____

_____

Name _____ Date _____

Plasma Cutter

Objective
Upon completion of this activity sheet, the student should be able to safely set up and operate a plasma cutter.

ASE Education Foundation Task Correlation

VI.B.1 Identify the considerations for cutting, removing, and welding various types of steel, aluminum, and other metals. **(HP-G)**

VI.B.3 Set up, attach work clamp (ground), and adjust the GMAW welder to "tune" for proper electrode stickout, voltage, polarity, flow rate, and wire-feed speed required for the substrate being welded. **(HP-I)**

We Support

ASE | Education Foundation

Tools and Equipment
Plasma cutter
Bare 2-inch × 2-inch sheet metal
Face shield
Saw horses or work bench

Safety Equipment
Safety glasses or goggles
Welding tinted face shield
Welding jacket
High-top safety shoes
Welding gloves

Introduction
In recent years plasma cutting has replaced oxyacetylene cutting in the collision field. It is neater, faster, and safer than traditional cutting.

Welder Description

Brand _____ Manufacturer _____

Procedure

Task Completed

1. Read the plasma cutter directions. Search related videos as to how to properly set up and use a plasma cutter. Familiarize yourself with the controls. ☐

2. Plug in the electrical line and air line. Put on the face shield. ☐

    a. What safety precautions must be taken before starting work? _____

    _____

    b. What voltage and air pressure are recommended when using the cutter? _____

    _____

3. Set a piece of sheet metal on saw horses. Place the work clamp on the sheet metal. Set the nozzle into the metal, and pull the trigger. Move fast enough to cut the metal without warping the panel. When you reach the end, release the trigger. ☐

4. Draw a 3-inch × 3-inch square on the panel. Cut the square out with the plasma cutter moving fast enough to cut it, yet not warp it. ☐

    a. Is there any warpage? If so, explain the cause. _____

    _____

    _____

5. Make a 24-inch long straight cut by clamping a straightedge to the sheet metal. The nozzle may be rested against the straightedge. Measure your cut. ☐

   a.   How much over or under is the cut? _____

6. Make a 2-inch diameter circular cutout with your plasma cutter. When complete, measure the diameter of the circle. ☐

   a.   How much over or under is your cut? _____

7. Lay a larger sheet under the original piece. Be sure to clamp them together. ☐

   a.   What will happen if the pieces are not clamped? _____

   _____

   b.   With both pieces clamped together, how thick are the combined pieces? _____

   c.   Can your plasma cutter cut through this thickness? _____. If so, cut the panel and check your results.

   d.   Is there any special procedure to shut the cutter down? If so, explain. _____

   _____

   _____

8. Shut down the plasma cutter, and store it in its proper location. ☐

INSTRUCTOR'S COMMENTS _____

_____

_____

Name _____ Date _____

Welding Aluminum

Objective

Upon completion of this activity sheet, the student should be able to safely weld aluminum with a MIG welder.

ASE Education Foundation Task Correlation

VI.B.1 Identify the considerations for cutting, removing, and welding various types of steel, aluminum, and other metals. **(HP-G)**

VI.B.2 Determine the correct GMAW welder type, electrode/wire type, diameter, and gas to be used in a specific welding situation. **(HP-I)**

VI.B.3 Set up, attach work clamp (ground), and adjust the GMAW welder to "tune" for proper electrode stickout, voltage, polarity, flow rate, and wire-feed speed required for the substrate being welded. **(HP-I)**

VI.B.5 Determine the proper angle of the gun to the joint and direction of gun travel for the type of weld being made. **(HP-G)**

VI.B.7 Identify hazards; foam coatings and flammable materials prior to welding/cutting procedures. **(HP-G)**

We Support

ASE | Education Foundation

Tools and Equipment

MIG welder
Argon shielding gas
#4043 electrode wire
Cable and clamp assembly
Manufacturer's or service manual
Appropriate hand tools as needed
¼-inch aluminum plate for welding
Tape measure

Safety Equipment

Welding gloves
Welding helmet
Welding jacket
Steel-toed shoes

Introduction

Many late model vehicles have a mixture of different plastic, steel, and aluminum parts. Great skill and caution must be used when welding aluminum.

Welder Description

Brand _____ Model _____

Procedure

Task Completed

1. Clean the weld area completely, both front and back, using wax and grease remover and a clean rag. ☐

   a. What is the purpose of the wax and grease remover? _____

   _____

Task Completed

b.   What will happen if the wax and grease remover is not used? _____

_____

c.   What is the purpose of cleaning both sides? _____

_____

2.   If the pieces to be welded are coated with a paint film, sand a strip about ¾-inch wide   ☐
to the bare metal, using a disc sander with #80 grit paper.

a.   Can the painted metal be welded over? _____

b.   What will happen to the metal piece if the paint is not sanded? _____

_____

3.   Do not press too hard, or the sander will heat up and peel off aluminum particles,   ☐
clogging the paper. Clean the metal with a stainless steel wire brush until it is shiny.

a.   Why is using a steel brush important? _____

_____

4.   Load 0.030 aluminum wire into the welder, and use 100 percent argon shielding gas.   ☐
Trigger to extend the wire about ¾ inch to 1 inch beyond the nozzle.

a.   How is the wire loaded in the welder? _____

_____

b.   When welding 0.030 wire, what is the recommended voltage? _____

c.   What is the recommended wire speed? _____

5.   Set the voltage, wire speed, and amount of argon gas according to the instructions   ☐
supplied with the welding machine. Remember that the wire speed must be faster than
for steel.

a.   Why must the wire speed be faster than for steel? _____

_____

6.   Set the tension of the wire drive roller lower to prevent twisting, but do not lower the   ☐
tension too much.

a.   What will happen if the tension is too loose? _____

_____

7.   Snip the end of the wire to remove the meltdown. Position the two pieces of aluminum   ☐
together. Hold the gun closer to the vertical when welding aluminum. Tilt only about
5–15 degrees from the vertical in the direction of the weld.

a.   Is this the push or pull technique? _____

8.   The entire joint will have a bead. Use only the forward welding method. Always push;
never pull. When making a vertical weld, start at the bottom and work up.

a.   How can you tell whether this is a push or pull technique? _____

_____

9.   Keep the distance between the contact tip and weld about ⁵⁄₁₆ inch to ⁹⁄₁₆ inch. This is   ☐
known as *stickout*.

10. If the arc is too large, turn down the voltage and increase the wire speed. The bead   ☐
should be uniform on top, with even penetration on the back. Is there any burn-through?

   a.   What does burn-through indicate? _____

      _____

   b.   How is burn-through rectified with machine settings? _____

      _____

INSTRUCTOR'S COMMENTS _____

_____

_____

Review Questions

Name _____ Date _____ Instructor Review _____

1. Technician A states that MIG welding is faster than stick welding. Technician B believes that MIG welding is easier to learn than stick welding. Who is correct?
   A. Technician A
   B. Technician B
   C. Both Technician A and Technician B
   D. Neither Technician A nor Technician B

2. MIG welding is acceptable on high-strength steel.
   A. True
   B. False

3. A sputtering sound without any arc means that the voltage is too _____.

4. Technician A says that MIG brazing is done with silicon bronze wire and 100 percent nitrogen. Technician B says it is done with silicon bronze wire and 100 percent argon. Who is correct?
   A. Technician A
   B. Technician B
   C. Both Technician A and Technician B
   D. Neither Technician A nor Technician B

5. The gun angle for the forehand and backhand methods is 10 to 30 degrees.
   A. True
   B. False

6. When using a squeeze-type resistance spot welder, both metal surfaces must be clamped tightly with no gap or it could cause poor current flow.
   A. True
   B. False

7. Technician A states that if the wire speed is set too slow, there will be spits and sputtering. Technician B believes that if the wire speed is too slow, there will be a much brighter reflected light. Who is correct?
   A. Technician A
   B. Technician B
   C. Both Technician A and Technician B
   D. Neither Technician A nor Technician B

8. Vertical welding is the technique of starting at the top and working your way downward.
   A. True
   B. False

9. When welding a butt joint, Technician A holds the gun at a 90-degree angle. Technician B holds the gun at a 30-degree angle. Who is correct?
   A. Technician A
   B. Technician B
   C. Both Technician A and Technician B
   D. Neither Technician A nor Technician B

10. Technician A makes a practice weld to test before he makes a structural weld. Technician B believes that a gun speed that is too slow will cause a melt-through. Who is correct?
    A. Technician A
    B. Technician B
    C. Both Technician A and Technician B
    D. Neither Technician A nor Technician B

Structural Glass Service

Name _____ Date _____ Instructor Review _____

Tool Identification

Identify and explain the proper use of each item or the method being used in the illustrations.

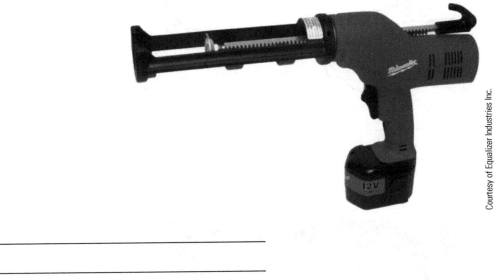

Courtesy of Equalizer Industries Inc.

_____

_____

_____

Courtesy of Equalizer Industries Inc.

_____

_____

_____

_____

_____

_____

_____

_____

_____

_____

_____

_____

_____

_____

_____

Name _____ Date _____

Remove and Install Structural Glass

Objective

Upon completion of this activity sheet, the student should be able to safely inspect, remove, and reinstall most urethane-bonded front windshields, back glass, and quarter panel glass.

ASE Education Foundation Task Correlation

I.D.1   Identify considerations for removal, handling, and installation of advanced glass systems (rain sensors, navigation, cameras, and collision avoidance systems). **(HP-G)**

I.D.2   Remove and reinstall or replace modular glass using recommended materials. **(HP-G)**

I.D.3   Check for water leaks, dust leaks, and wind noise. **(HP-G)**

We Support

ASE | **Education Foundation**

Tools and Equipment

Vehicle
Windshield removal tools (vary from vehicle to vehicle)
Appropriate sockets and hand tools
Windshield urethane cartridge
Urethane cartridge dispenser gun
Friend to help carry and set windshield
Flashlight
Water hose without the spray nozzle
Paper towels
Rubbing alcohol

Safety Equipment

Work gloves
Safety glasses or goggles

Introduction

Automotive windshields are considered part of the vehicle structure, even though most people never think of the glass that seriously. Structural glass makes up 60 percent of the vehicle's structural integrity. That is a staggering number to think about being so responsible for the safety of the vehicle structure. Making sure that you use the correct adhesives and procedures is extremely critical to ensure the strength and integrity of the glass holds up in a collision. This job sheet will walk you through how to remove most urethane-bonded front windshields, back glass, and quarter panel glass.

Vehicle Description

Year _____ Make _____ Model _____

VIN _____

Procedure

Task Completed

1. First, you must start with removing the trim around the exterior of the windshield. This includes the windshield wiper arms and the wiper cowl panel. Sometimes there are side pillar trim that must be removed on trucks and SUVs when you open the front doors.   ☐

2. Disconnect any wires running to the rearview mirror on the inside of the glass. Remove the rearview mirror from the glass.   ☐

3. Remove the A pillar trim pieces that run up the windshield posts on the inside of the vehicle. Be extremely careful when removing any clips, fasteners, or retainers. Avoid breaking or losing them because they can usually be reused.   ☐

4. Time to cut out the old windshield using a windshield removal tool. This can be done by hand or by using a pneumatic windshield removal tool for faster removal. The goal is to not be too fast and crack or shatter the windshield and cause damage to the pinch weld or the painted panels around the windshield.   ☐

5. The easiest and least expensive way to remove the windshield is to use an 18-inch urethane cutout knife, which is typically used by professionals. Cut the urethane a round the entire perimeter of the windshield. You may have to trim the urethane around the perimeter of the windshield a couple of times in a few places to safely remove the urethane and not damage the pinch welds.

6. Once the glass is out, if it is to be reused once the finished process has been completed, you must use a heavy duty glass scraper blade to remove all the old urethane from the windshield.   ☐

7. You can use the same knife or blade to trim down the urethane on the pinch weld on the body to a smooth and level surface. Remove all loose flaps or fragments of windshield urethane that are not properly sealed down. When you reapply the windshield urethane to itself during the installation process, these flaps will cause improper adhesion and allow moisture and water to get in the vehicle.   ☐

NOTE: It may help to do a quick Google search or YouTube video search on glass removal and installation on a certified YouTube channel. Safelite Auto Glass has a large inventory of instructional videos.

8. Now that the windshield urethane has been trimmed, clean any dirt or debris left on the frame. There should be no dirt or debris where the urethane is going to be applied. Even if the windshield is new, you should always use a proper glass cleaner before applying any primer or urethane.   ☐

9. Prime the pitch weld. As you cut out the old urethane, you may have made some scratches along the pinch weld. Apply a urethane windshield primer to the perimeter of the windshield, inside edge, and in the window channel, making sure to cover up all scratches.   ☐

10. The primer can help the urethane bond better and act to prohibit rust when applied to the window channel, especially if there are scratches. Always use urethane primer to ensure that rust does not develop. If there is rust in this area, you must sand it down and then apply primer. After you have applied the primer, wait the recommended time before you apply the urethane bead.   ☐

NOTE: If the old windshield urethane has been trimmed down smoothly and still sealed to the glass, with no tears, flaps, or punctures, you may apply the new urethane on top of the old urethane. The same goes for the old urethane on the window channel of the body.

11. Apply the urethane. Start by removing the bottom cap from the urethane cartridge, and then install it on your caulking gun. If you trim the cartridge tip in the shape of a "V," it will allow for a taller, larger surface of a urethane bead to come out when dispensed. This gives more area for the windshield urethane to mash down and spread out to make a nice even seal when the windshield is installed.   ☐

12. Carefully lay the bead of urethane, and try to keep it in a continuous line. Avoid stops □
and make sure to move your hands at a smooth and constantly steady speed. If you
miss any spots or the bead is too thin, just come back later after you have completed
the full circle. If there are any spots where it is not continuous, you can apply a small
amount of urethane, and use a paint stick or a plastic windshield tool to fix the urethane.

13. Be careful not to let any urethane fall on the vehicle's paint or trim. Removing it can be □
very exhausting; use paper towels and rubbing alcohol if any urethane falls on the car's
paint or trim.

14. Next, install the new windshield. I recommend you get a second person to help you with □
this step. Have your helper assist you in placing the windshield and the proper location
in the vehicle. Once the windshield is placed, it is OK to shift it to its proper position, but
keep the movement of the windshield to a minimum.

15. Once finished, you may want to apply two strips of 2-inch-wide masking tape to the □
surface of the windshield and the roof of the vehicle about 4–6 inches in length. This
will help hold it in place for the next couple of hours until fully cured.

Water Leak Test
Procedure

16. Once the urethane has had a chance to dry for 30 minutes to an hour, you can perform □
a water leak test.

17. You must get inside the vehicle with a flashlight and some paper towels just in case. □
Have your friend run water over the glass and surrounding area from a water hose
without a spray nozzle on it.

   IMPORTANT: DO NOT use a spray nozzle for this test.

   a.   What would happen if you did use a spray nozzle on the water hose for this
        exercise?

   _____

   _____

18. If you do not see any water running in or dripping, you have made a proper windshield □
replacement. If water is running or dripping inside, well, it is not going to be a fun time
taking this back out and starting over. But that is what is going to have to happen.

19. Once there are no leaks, you may begin to reinstall all the exterior and interior trim. □
panels Make sure to reattach the rearview mirror and plug up any wiring harness that
may run to the mirror itself.

INSTRUCTOR'S COMMENTS _____

_____

_____

Review Questions

Name _____ Date _____ Instructor Review _____

1. _____ consists of two thin sheets of glass.

2. _____ is a transparent substance manufactured by heating a mixture of sand, soda (sodium carbonate), limestone, and other materials.

3. Tempered glass is always used for windshields.
   A. True
   B. False

4. _____ glass provides added protection against shattering and cuts during impact.

5. The _____ method requires all of the old adhesive to be cut out.

6. Technician A says to scrub back and forth when removing glass particles from carpet. Technician B says to turn all the vents on when vacuuming broken glass particles out of the automobile. Who is correct?
   A. Technician A
   B. Technician B
   C. Both Technician A and Technician B
   D. Neither Technician A nor Technician B

7. Technician A says that temperature and humidity play a large factor in the curing process of windshield urethane. Technician B says that some windshield urethanes may take up to 24 hours to be fully cured. Who is correct?
   A. Technician A
   B. Technician B
   C. Both Technician A and Technician B
   D. Neither Technician A nor Technician B

8. Technician A says when removing a glass windshield, you may use a heat gun to soften up the adhesive. Technician B says that a hot knife can be used to cut through the adhesive also. Who is correct?
   A. Technician A
   B. Technician B
   C. Both Technician A and Technician B
   D. Neither Technician A nor Technician B

9. Technician A says that pinch weld primer will not bond to bare metal. Technician B says pinch weld primer will bond to automotive primer. Who is correct?
   A. Technician A
   B. Technician B
   C. Both Technician A and Technician B
   D. Neither Technician A nor Technician B

10. A quarter panel was replaced and a few plug welds were not fully welded and have pits left behind. Technician A says this is OK because the windshield urethane will fill them in. Technician B says water will run under the adhesive and into the passenger compartment. Who is correct?
    A. Technician A
    B. Technician B
    C. Both Technician A and Technician B
    D. Neither Technician A nor Technician B

Body/Frame Damage Measurement

Name _____ Date _____ Instructor Review _____

Identifying Conventional Frame Damage

Next to each figure shown here, identify the type of frame damage.

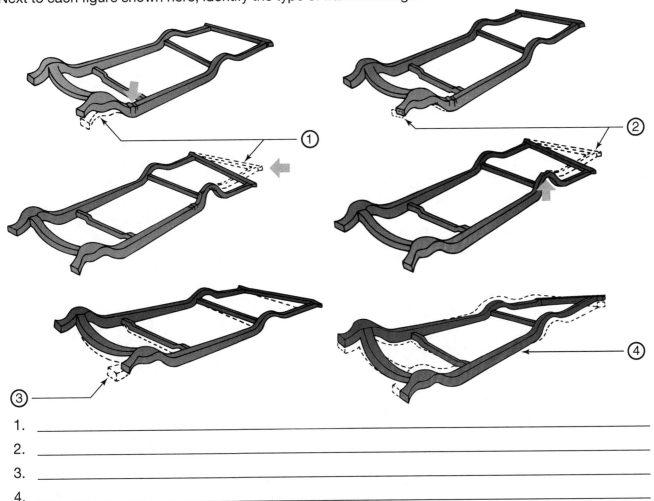

1. _____

2. _____

3. _____

4. _____

Name _____ Date _____ Instructor Review _____

Simulated Vehicle Measurements

1. X Check Engine Compartment

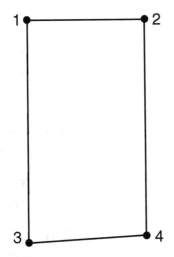

1–4   mm   _____

      inches _____

2–3   mm   _____

      inches _____

2. Underhood Measurements

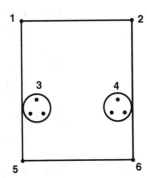

Length

1–5   mm   _____
      inches _____

2–6   mm   _____
      inches _____

Width

1–2   mm   _____
      inches _____

3–4   mm   _____
      inches _____

5–6   mm   _____
      inches _____

X checks

1–4   mm   _____
      inches _____

2–3   mm   _____
      inches _____

3–6   mm   _____
      inches _____

1–6   mm   _____
      inches _____

2–5   mm   _____
      inches _____

3. Radius Measurement

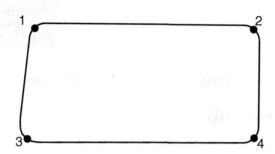

X checks

1–4	mm	_____
	inches	_____
2–3	mm	_____
	inches	_____

Draw straight lines to form a point; measure point to point.

4. Underbody Measurements—Full-Frame Pickup Truck

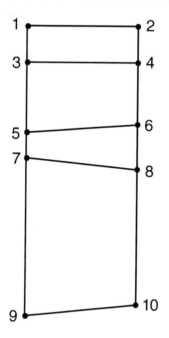

X checks

1–6	mm	_____
	inches	_____
2–5	mm	_____
	inches	_____
1–4	mm	_____
	inches	_____
2–3	mm	_____
	inches	_____
5–8	mm	_____
	inches	_____
6–7	mm	_____
	inches	_____
7–10	mm	_____
	inches	_____
8–9	mm	_____
	inches	_____

5. Upper Body Measurements—Door Opening

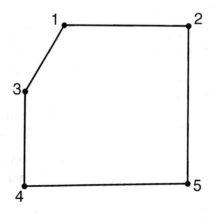

X checks

1–5	mm	_____
	inches	_____
3–5	mm	_____
	inches	_____
2–4	mm	_____
	inches	_____

6. Deck Lid Opening

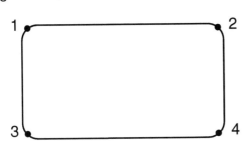

X checks

1–4 mm _____
inches _____

2–3 mm _____
inches _____

7. Windshield Opening

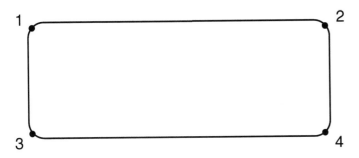

X checks

1–4 mm _____
inches _____

2–3 mm _____
inches _____

Name _____ Date _____ Instructor Review _____

Misalignment Measurements

1. Write in these measurements:

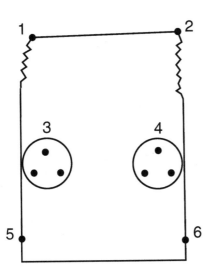

Width

1–2 _____

3–4 _____

7–8 _____

X checks

1–6 _____

2–5 _____

1–8 _____

2–7 _____

What types of damage conditions are indicated? _____

_____

_____

2.

Width

1–2 _____

5–6 _____

X checks

1–4 _____

2–3 _____

1–6 _____

2–5 _____

What types of damage conditions are indicated? _____

_____

_____

3.

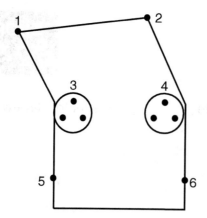

Width

1–2 _____

5–6 _____

X checks

1–4 _____

2–3 _____

1–6 _____

2–5 _____

What types of damage conditions are indicated? _____

_____

_____

Name _____ Date _____

▶Unibody Measurements

Objective

Upon completion of this activity sheet, the student should be able to make measurements on any vehicle and compare them to reference sheets.

ASE Education Foundation Task Correlation

I.A.1   Select and use proper personal safety equipment; take necessary precautions with hazardous operations and materials in accordance with federal, state, and local regulations. **(HP-I)**

I.A.2   Locate procedures and precautions that may apply to the vehicle being repaired. **(HP-I)**

I.C.3   Measure and diagnose unibody damage using tram gauge. **(HP-I)**

I.C.4   Measure and diagnose unibody vehicles using a dedicated (fixture) measuring system. **(HP-G)**

I.C.5   Diagnose and measure unibody vehicles using a three-dimensional measuring system (mechanical, electronic, and laser, etc.). **(HP-I)**

We Support

ASE | **Education Foundation**

Tools and Equipment

Undamaged unibody car
Tape measure
Frame dimension guide
Tram gauges

Safety Equipment

Safety glasses or goggles

Introduction

With more and more unibody vehicles being produced today, the straightening of frames is more critical than ever. Incorrect pulls can alter the function of important components, causing damage or injury. Precise measurements will allow new parts to line up to factory specifications.

Vehicle Description

Year _____ Make _____ Model _____

VIN _____

Procedure

Task Completed

1. Use a tram gauge to cross-check the engine compartment. Consult the frame dimension guide for points to measure. Make sure that the gauges are as level as possible and freely sitting on the measurement locations. Do not bend them or force them.   ☐

   a.   What points are you measuring from and to? _____

   b.   What is the actual measurement? _____

   c.   What is the reference measurement? _____

   d.   If there is a difference, what is it? _____

   _____

**Task
Completed**

2. Use a tram to measure a door opening. ☐

  a. What are your reference measuring points? _____

  b. What is the actual measurement? _____

  c. What is the reference measurement? _____

  d. If there is a difference, what is it? _____

3. Use a tram to measure the windshield opening. ☐

  a. What are your reference measuring points? _____

  b. What is the actual measurement? _____

  c. What is the reference measurement? _____

  d. If there is a difference, what is it? _____

Using a different undamaged unibody car, fill in the following information.

4. Measure from the upper rear bolt on the left fender to the upper front bolt on the
    right fender. ☐

    Measure from the upper rear bolt on the right fender to the upper front bolt on the
    left fender. ☐

  a. Are these measurements the same? _____

  b. If not, what is the difference? _____

  c. Which side is longer? _____

  d. What is the width between the fenders at the front? _____

  e. What is the width between the fenders at the rear? _____

5. Measure from the left shock tower to the center of the hood latch hole. ☐

    Measure from the right shock tower to the center of the hood latch hole. ☐

  a. Are these measurements the same? _____

  b. If not, what is the difference? _____

  c. Which side is longer? _____

  d. What is the width between the shock towers? _____

6. On the right door, measure the distance between the striker and the top fender hinge. ☐

    Measure the distance between the striker and bottom hinge. ☐

    On the left door, measure the distance between the striker and top fender hinge. ☐

    Measure the distance between the striker and bottom hinge. ☐

  a. Is there any difference between the left and right door? _____

  b. If so, what is the difference? _____

  c. According to your reference, which door is off? By how much? _____

      _____

INSTRUCTOR'S COMMENTS _____

_____

_____

Name _____ Date _____

Damaged Vehicle Measurements

Objective

Upon completion of this activity sheet, the student should be able to accurately make measurements on a damaged vehicle.

ASE Education Foundation Task Correlation

I.A.1	Select and use proper personal safety equipment; take necessary precautions with hazardous operations and materials in accordance with federal, state, and local regulations. **(HP-I)**
I.A.2	Locate procedures and precautions that may apply to the vehicle being repaired. **(HP-I)**
I.C.1	Analyze and identify misaligned or damaged steering, suspension, and powertrain mounting points that can cause vibration, steering, and chassis alignment problems. **(HP-G)**
I.C.3	Measure and diagnose unibody damage using tram gauge. **(HP-I)**
I.C.4	Measure and diagnose unibody vehicles using a dedicated (fixture) measuring system. **(HP-G)**
I.C.5	Diagnose and measure unibody vehicles using a three-dimensional measuring system (mechanical, electronic, and laser, etc.). **(HP-G)**
I.C.6	Determine the extent of the direct and indirect damage and the direction of impact; plan and document the methods and sequence of repair. **(HP-I)**

We Support

ASE | **Education Foundation**

Tools and Equipment

Damaged unibody car
Tape measure
Tram gauge

Safety Equipment

Safety glasses or goggles

Introduction

Accurate measurements are as important as the actual pulling of the damaged parts. If accurate measurements are not taken, more damage could be caused, leading to a dangerous driving condition.

Vehicle Description

Year _____ Make _____ Model _____

VIN _____

Procedure

Task Completed

Walk around a damaged unibody car. Note all the damage. List the improper gaps next.

1. Measure the following gaps to the closest 1/16 inch.   ☐

   a.   Hood to right fender. Factory measurement versus actual. _____

   If there is a difference, how would you correct it? _____

   _____

   _____

b. Right fender to front door. Factory measurement versus actual. _____

If there is a difference, how would you correct it? _____

_____

_____

c. R/F door to R/R door. Factory measurement versus actual. _____

If there is a difference, how would you correct it? _____

_____

_____

d. R/R door to R/R quarter panel. Factory measurement versus actual. _____

If there is a difference, how would you correct it? _____

_____

_____

e. R/R quarter panel to deck lid. Factory measurement versus actual. _____

If there is a difference, how would you correct it? _____

_____

_____

f. Deck lid to L/R quarter panel. Factory measurement versus actual. _____

If there is a difference, how would you correct it? _____

_____

_____

g. L/R quarter panel to L/R door. Factory measurement versus actual. _____

If there is a difference, how would you correct it? _____

_____

_____

h. L/R door to L/F door. Factory measurement versus actual. _____

If there is a difference, how would you correct it? _____

_____

_____

i. L/F door to L/F fender. Factory measurement versus actual. _____

If there is a difference, how would you correct it? _____

_____

_____

j.   L/F fender to hood. Factory measurement versus actual. _____

    If there is a difference, how would you correct it? _____

_____

_____

2.   List the damaged major outer body panels from the accident and whether they need to   ☐
be replaced or not. _____

_____

_____

3.   Measure from the center of the left front wheel to the center of the left back wheel.   ☐
Factory measurements versus actual. _____

4.   Measure from the center of the right front wheel to the center of the right rear wheel.   ☐
Factory measurements versus actual. _____

    a.   Is there any difference between the right and left side? _____

_____

INSTRUCTOR'S COMMENTS _____

_____

_____

Review Questions

Name _____ Date _____ Instructor Review _____

1. _____ are thick metal arms that bolt between the vehicle and the bench.

2. A _____ uses beams of light to take measurements.

3. A _____ uses large steel fixtures to determine whether the frame or unibody has been pushed out of alignment in a collision.

4. Crush zones may be holes, dimples, or slots that are formed into a part.
   A. True
   B. False

5. A three-dimensional laser system measures length, width, and height.
   A. True
   B. False

6. Making pulls with a chain "bridge" takes more than one chain.
   A. True
   B. False

7. Technician A says that a tram can be used to measure a door opening; Technician B says that a tram cannot be used to measure a door opening. Who is correct?
   A. Technician A
   B. Technician B
   C. Both Technician A and Technician B
   D. Neither Technician A nor Technician B

8. Technician A says that sonic systems and robot arm systems are the main measuring systems used. Technician B says that sonic and laser measuring are the most used. Who is correct?
   A. Technician A
   B. Technician B
   C. Both Technician A and Technician B
   D. Neither Technician A nor Technician B

9. Technician A says that all laser systems use the metric system. Technician B says that all systems use the English measuring system. Who is correct?
   A. Technician A
   B. Technician B
   C. Both Technician A and Technician B
   D. Neither Technician A nor Technician B

10. Technician A says that the rear axle sits in the "cradle." Technician B says that the engine sits in the cradle. Who is correct?
    A. Technician A
    B. Technician B
    C. Both Technician A and Technician B
    D. Neither Technician A nor Technician B

Unibody/Frame Realignment

Name _____ Date _____ Instructor Review _____

Clamp Identification

Fill in the proper names of the pulling clamps or plates shown in the figure that follows.

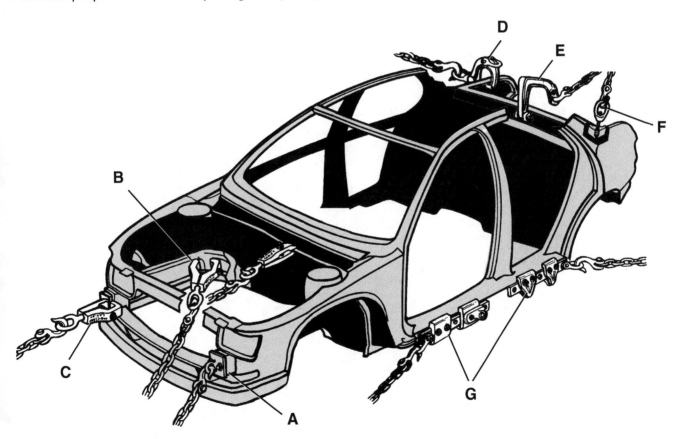

A. _____

B. _____

C. _____

D. _____

E. _____

F. _____

G. _____

Name _____ Date _____ Instructor Review _____

Pull Plans

In this example, the center section and right side need to be held in place. The left corner needs to be moved sideways and down.

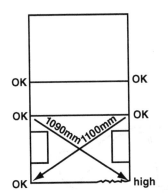

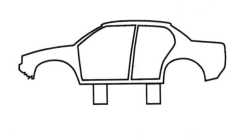

This drawing indicates the hookups and tie-downs needed to repair this vehicle.

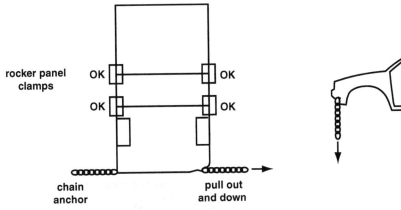

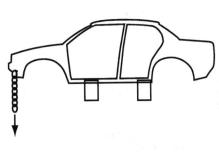

Draw in the hookups and tie-downs on the following damage diagrams.

1.

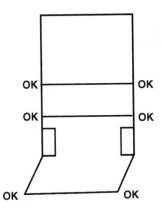

2.

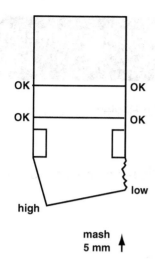

mash
5 mm

roof damage

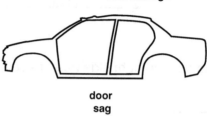

door
sag

3.

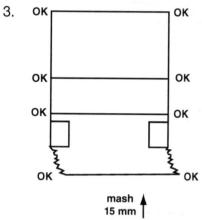

mash
15 mm

4.

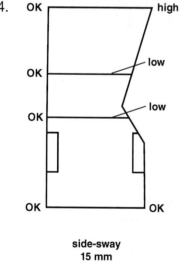

side-sway
15 mm

roof damage

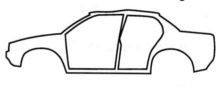

center pillow damage
rocker panel damage

5.

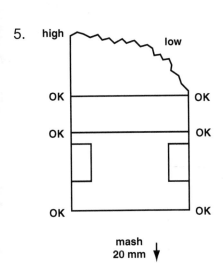

mash
20 mm ↓

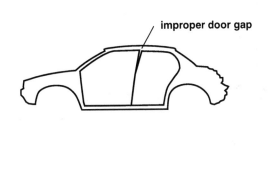

improper door gap

6.

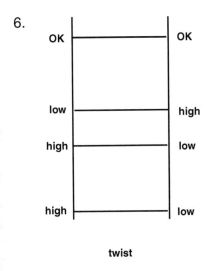

twist

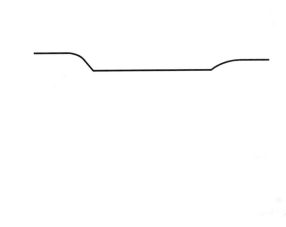

7.

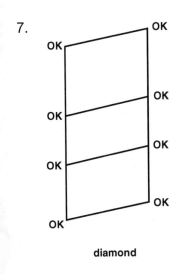

diamond

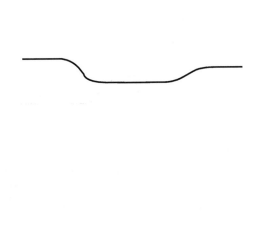

8.

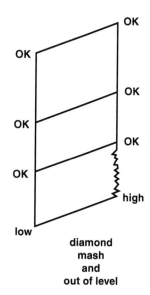

diamond
mash
and
out of level

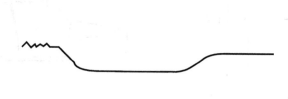

9.

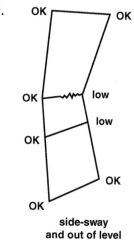

side-sway
and out of level

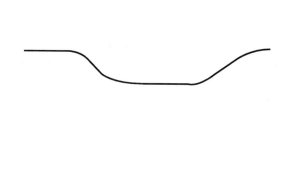

10.

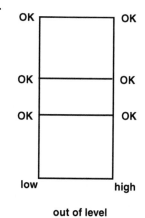

out of level

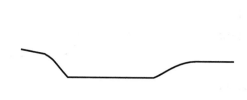

Name _____ Date _____ Instructor Review _____

Identifying Pulls

In the figure that follows, use an arrow to show the direction of the pull and describe what kind
of pull it is.

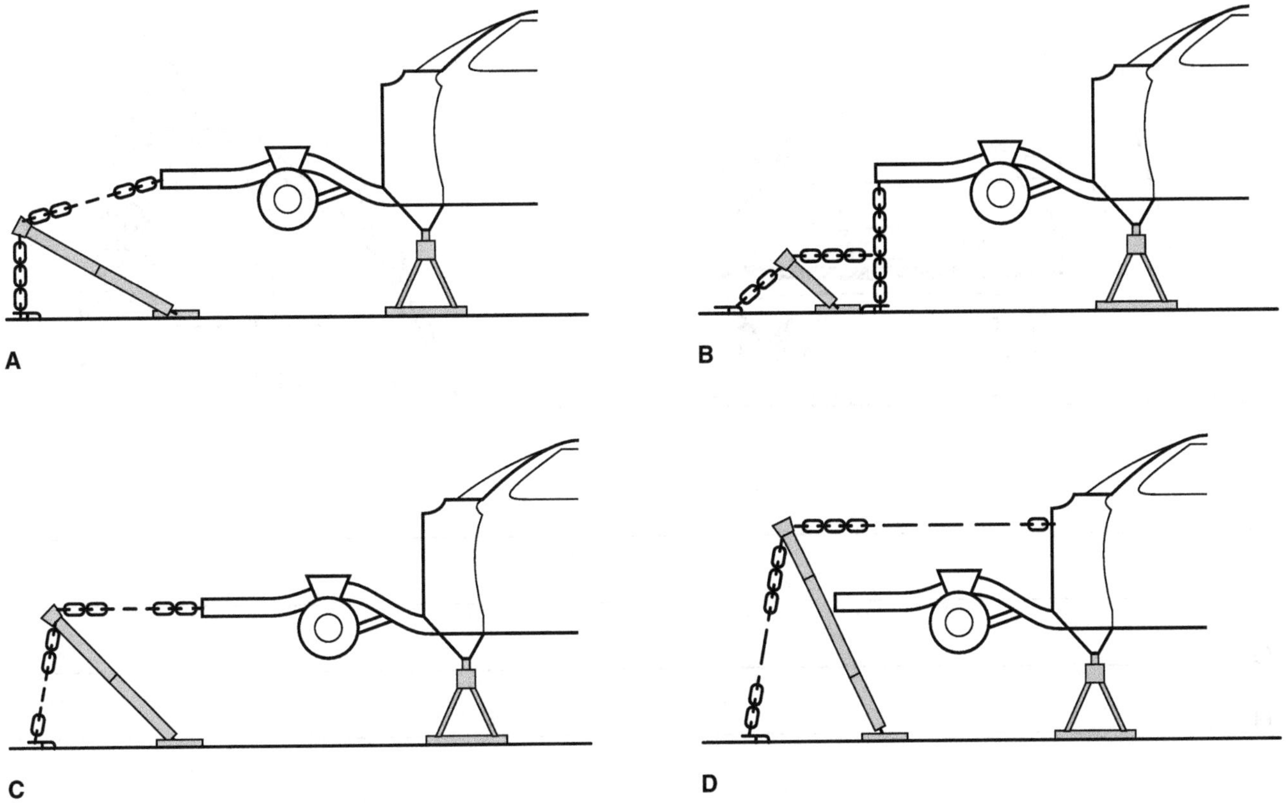

A

B

C

D

A. _____

B. _____

C. _____

D. _____

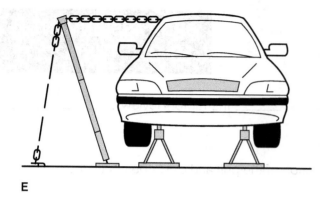

E

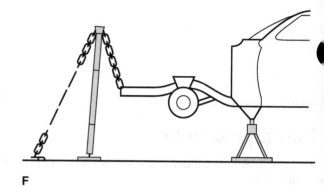

F

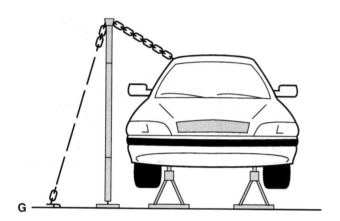

G

H

E. _____

F. _____

G. _____

H. _____

Name _____ Date _____

Unibody Mounting and Analysis

Objective

Upon completion of this activity sheet, the student should be able to safely mount a unibody car on a bench or frame rack. The student should also be able to analyze and diagnose the damage and types of damage present.

ASE Education Foundation Task Correlation

I.A.1	Select and use proper personal safety equipment; take necessary precautions with hazardous operations and materials in accordance with federal, state, and local regulations. **(HP-I)**
I.A.2	Locate procedures and precautions that may apply to the vehicle being repaired. **(HP-I)**
I.A.3	Identify vehicle system hazard types (supplemental restraint system [SRS], hybrid/electric/alternative fuel vehicles), locations, and recommended procedures before inspecting or replacing components. **(HP-I)**
I.C.3	Measure and diagnose unibody damage using tram gauge. **(HP-I)**
I.C.6	Determine the extent of the direct and indirect damage and the direction of impact; plan and document the methods and sequence of repair. **(HP-I)**
I.C.7	Attach anchoring devices to vehicle; remove or reposition components as necessary. **(HP-I)**
I.C.13	Identify substrate and repair or replacement recommendations. **(HP-I)**
I.C.14	Identify proper cold stress relief methods. **(HP-I)**
I.C.15	Repair damage using power tools and hand tools to restore proper contours and dimensions. **(HP-I)**
I.C.16	Determine sectioning procedures of a steel body structure. **(HP-I)**
I.C.17	Remove and replace damaged structural components. **(HP-G)**
I.C.18	Restore corrosion protection to repaired or replaced structural areas and anchoring locations. **(HP-I)**
I.C.19	Determine the extent of damage to aluminum structural components; repair, weld, or replace. **(HP-G)**
I.C.20	Analyze and identify crush/collapse zones. **(HP-I)**

We Support
Education Foundation

Tools and Equipment

Unibody car
Tape measure
Frame dimension guide
Tram gauge

Safety Equipment

Safety glasses or goggles
Work gloves

Introduction

With the prevalence of unibody vehicles, understanding how to make quality repairs to unibody vehicles is critical. Precise measurements will allow new parts to line up to factory specifications.

Vehicle Description

Year _____ Make _____ Model _____

VIN _____

Procedure

1. Position the vehicle directly in front of the frame rack. Drive it up onto the rack, or if it is not able to start, attach a winch to the frame rack and pull the vehicle up onto it.   ☐

2. Once on the rack, put it in park and pull the emergency brake so that it does not roll. Next, look underneath for the proper locations of where you can place the pinch weld clamps without clamping shut any drain holes or hitting any fuel line hoses or brake lines.   ☐

3. Place blocks or steel bars underneath to help avoid damaging exhaust, transmission, or other parts under the car while lifting the car up by the center lift in the rack.   ☐

4. Once the car is lifted high enough to put the pinch weld clamps under the car, slide the clamps under and line them up so that they will be in the proper location. Make sure they are not too far forward and the car tilts to the front and also not too far rearward so that it does not tilt rearward. The clamps should be separated as far to the front of the middle section and as far to the rear of the middle section as you can to safely attach them to t he vehicle, so it will balance and distribute the load evenly.   ☐

    a.   Are there any obstructions in the way of attaching the clamps where they should go? If so, how will you remove them or adjust the clamps to fit? _____
    _____

5. Slowly lower the car down to the pinch weld clamps. Make sure the clamps do not crush a brake line, cable, wiring harness, or gas line.   ☐

    a.   What are some of the dangers from a crushed brake line, cable, wiring harness, or gas line?
    _____

6. Use an impact wrench to tighten the clamp bolts. When the car is securely mounted to the pinch weld clamps, tighten down all of the bolts from the pinch weld clamps to the frame rack itself. Go back over **ALL** bolts in this process to ensure they are all tight and secure. This is a very important step, so please **DO NOT** overlook it.   ☐

    a.   What size socket or sockets did you use to tighten the clamps? _____
    _____

    b.   Is there a sequence in the pattern of the clamps? _____
    _____

The next steps in this job sheet are for a damaged vehicle needing measuring, analysis, and repair.

7. Now that it is clamped into position, it is time to measure the damage so that we know how the damage happened and how to properly plan to correct it.   ☐

8. List the direct damage (or primary damage) on the vehicle. _____
_____
_____

9. List the indirect damage on the vehicle. _____
_____
_____

10. Get the body and frame measurements by whatever software or frame dimensions you have.   ☐

11. Measure the damaged area by the actual measurements that are listed on the frame specification sheets or software screen.

    a.   Judging by the damage, in what direction and/or angle do you believe the car was hit?

    _____

    _____

    b.   Are any crush zones or collapse zones damaged? If so, please describe.

    _____

    _____

12. If repairs are needed, what panels would need to be repaired, and what substrate are they?

_____

_____

_____

_____

13. What panels would need to be replaced, and what substrate are they?

_____

_____

_____

    a.   If any panels have to be sectioned to be replaced instead of full panel replacement, list what type of substrate the panel is, and how would you determine the sectioning procedures for this task. _____

    _____

    _____

14. When repairing frame damage, how do you cold stress relieve damaged areas?

_____

_____

15. List the steps you will take or you did take to repair the damaged structural areas and the tools that were used. _____

_____

_____

_____

_____

_____

_____

INSTRUCTOR'S COMMENTS _____

_____

_____

Name _____ Date _____

Body-Over-Frame Mounting and Analysis

Objective

Upon completion of this activity sheet, the student should be able to safely mount a body-over-frame type automobile on a bench or frame rack. Plus be able to analyze and diagnose the damage and types of damage present.

ASE Education Foundation Task Correlation

I.A.1   Select and use proper personal safety equipment; take necessary precautions with hazardous operations and materials in accordance with federal, state, and local regulations. **(HP-I)**

I.A.2   Locate procedures and precautions that may apply to the vehicle being repaired. **(HP-I)**

I.A.3   Identify vehicle system hazard types (supplemental restraint system [SRS], hybrid/electric/alternative fuel vehicles), locations, and recommended procedures before inspecting or replacing components. **(HP-I)**

I.B.1   Measure and diagnose structural damage using a tram gauge. **(HP-I)**

I.B.2   Attach vehicle to anchoring devices. **(HP-G)**

I.B.8   Remove and replace damaged structural components. **(HP-G)**

I.B.15   Determine the extent of the direct and indirect damage and the direction of impact; document the methods and sequence of repair. **(HP-I)**

I.B.16   Analyze and identify crush/collapse zones. **(HP-I)**

We Support

Education Foundation

Tools and Equipment

Truck or SUV with body over frame
Tape measure
Frame dimension guide
Tram gauge

Safety Equipment

Safety glasses or goggles
Work gloves

Introduction

With the body of the vehicle mounting directly over the frame, understanding how to make quality repairs to body-over-frame type vehicles is critical. Precise measurements will allow new parts to line up to factory specifications.

Vehicle Description

Year _____ Make _____ Model _____

VIN _____

Task Completed

Procedure

1. Position the vehicle directly in front of the frame rack. Drive it up onto the rack, or if it is not able to start, attach a winch to the frame rack and pull the vehicle up onto it.   ☐

2. Once on the rack, put it in park and pull the emergency brake so that it does not roll. ☐
   Next, look underneath for the proper locations of where you can place the frame
   attachment clamps without hitting any crossmembers, fuel line hoses, or brake lines.

3. Place any blocks or steel bars underneath to help avoid damaging any exhaust, ☐
   transmission, or other parts under the frame while lifting the vehicle up by the center
   lift in the rack.

4. Once the vehicle is lifted high enough to put the frame attachment clamps under the ☐
   vehicle, slide the clamps under and line them up so that they will be in the proper
   location. Make sure they are not too far forward and the car tilts to the front and also
   not too far rearward so that it does not tilt rearward. The clamps should be separated
   as far to the front of the middle section and as far to the back of the middle section as
   you can to safely attach them to the vehicle so it will balance and distribute the load
   evenly.

   a.   Are there any obstructions in the way of attaching the clamps where they should
        go? If so, how will you remove them or adjust the clamps to fit?

   _____

5. Slowly lower the car down to the frame attachment clamps. Make sure the clamps do ☐
   not crush a brake line, cable, wiring harness, or gas line.

   a.   What are some of the dangers from a brake line, cable, wiring harness, or gas line?

   _____

   _____

6. Use an impact wrench to tighten the clamp bolts. When the car is securely mounted to ☐
   the frame attachment clamps, tighten down all of the bolts from the frame attachment
   clamps to the frame rack itself. Go back over **ALL** bolts in this process to ensure they
   are all tight and secure. This is a very important step, so please **DO NOT** overlook it.

   a.   What size socket or sockets did you use to tighten the clamps? _____

   b.   Is there a sequence in the pattern of the clamps? _____

   _____

   **The next steps in this job sheet are for a damaged vehicle needing measuring,
   analysis, and repair.**

7. Now that it is clamped into position, it is time to measure the damage so that we know ☐
   how the damage happened and how to properly plan to correct it.

8. List the direct (or primary) damage on the vehicle. _____

   _____

   _____

   _____

9. List the indirect damage on the vehicle. _____

   _____

   _____

   _____

10. Get the body and frame measurements by whatever software or frame dimensions ☐
    you have.

11. Measure the damaged area and beyond the damaged area by the actual measurements ☐
that are listed on the frame specification sheets or software screen so you can tell where
the damage is and where it starts.

   a. Judging by the damage, what direction and/or angle do you believe the car
   was hit? _____

   _____

   b. Are any crush zones or collapse zones damaged? If so, please describe.

   _____

   _____

12. If repairs are to be done now, what panels would need to be repaired, and what
substrate are they? _____

_____

13. What panels would need to be replaced, and what substrate are they? _____

_____

   a. If any panels have to be sectioned to be replaced instead of full panel replacement,
   list what type of substrate the panel is, and how would you determine the sectioning
   procedures for this task. _____

   _____

14. When repairing frame damage, how do you cold stress relieve damaged areas?

_____

15. List the steps of how you will repair or did repair the damaged areas and the tools used.

_____

_____

_____

_____

INSTRUCTOR'S COMMENTS _____

_____

_____

Name _____ Date _____

Universal Bench Mounting

Objective
Upon completion of this activity sheet, the student should be able to safely mount a unibody car on a bench.

ASE Education Foundation Task Correlation

I.A.1   Select and use proper personal safety equipment; take necessary precautions with hazardous operations and materials in accordance with federal, state, and local regulations. **(HP-I)**

I.A.2   Locate procedures and precautions that may apply to the vehicle being repaired. **(HP-I)**

I.A.3   Identify vehicle system hazard types (supplemental restraint system [SRS], hybrid/electric/alternative fuel vehicles), locations, and recommended procedures before inspecting or replacing components. **(HP-I)**

I.C.7   Attach anchoring devices to vehicle; remove or reposition components as necessary. **(HP-I)**

We Support

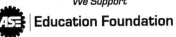

ASE | **Education Foundation**

Tools and Equipment
Undamaged unibody car
Tape measure
Frame dimension guide
Tram gauge

Safety Equipment
Safety glasses or goggles

Introduction
With the prevalence of unibody vehicles, understanding how to make quality repairs to unibody vehicles is critical. Precise measurements will allow new parts to line up to factory specifications.

Vehicle Description

Year _____ Make _____ Model _____

VIN _____

Procedure

Task Completed

1. Read the instruction guide on mounting the vehicle. Some benches are assembled under the car. In other cases, the car is raised up and the bench is rolled under the car. These instructions will be for the latter case.   ☐

   a.   What type of bench does your shop have? _____

   b.   Who is the manufacturer? _____

2. Safely raise the vehicle to bench clearance height.   ☐

   a.   How is the car raised? _____

   _____

   b.   What is the bench clearance height? _____

   c.   What are some of the safety concerns? _____

   _____

3. A properly designed lift is always recommended. Some manufacturers use a flow jack   ☐
    and wheel stands. Roll the bench under the car. Line up the pinch weld clamps so that
    they will be in the proper location.

    a.   What is the proper location of the pinch weld clamps? _____

    _____

4. Lock the clamps in place.   ☐

    a.   How are the clamps locked? _____

    _____

5. Slowly lower the car down to the pinch weld clamps. Make sure the clamps do not   ☐
    crush a brake or gas line.

    a.   What are some of the dangers from a crushed brake or gas line? _____

    _____

6. Use an impact wrench to tighten the clamp bolts. When the car is securely mounted   ☐
    on the bench, the bench and the car may be moved.

    a.   What size socket did you use to tighten the clamps? _____

    b.   Is there a sequence in the pattern of the clamps? If so, what is the sequence? ____

    _____

7. Consult the manual provided by the bench manufacturer to determine locations to   ☐
    mount fixtures, pins, or targets.

    a.   Where are these locations on the vehicle? _____

    _____

8. Any fixture or pin that does not reach the intended point indicates damage.

    a.   Are there any parts that do not line up to their designated points? _____

    b.   If so, describe the damage. _____

    _____

    c.   Is there more than one section that does not line up? _____

    d.   If so, what does that indicate? _____

    _____

INSTRUCTOR'S COMMENTS _____

_____

_____

Review Questions

Name _____ Date _____ Instructor Review _____

1. When pulling a vehicle, a pull force or _____ should be applied.

2. The force of all pulls combined is also known as the _____.

3. Lateral deflection is _____ _____ damage.

4. Traction direction refers to the direction of the pulling force applied.
   A. True
   B. False

5. All floor straightening systems have anchor pots or rails.
   A. True
   B. False

6. Pinch weld clamps are used to anchor vehicles with full frames.
   A. True
   B. False

7. Technician A says that pulling clamps are attached to the vehicle. Technician B says that the pulling clamps are attached to the frame bench. Who is correct?
   A. Technician A
   B. Technician B
   C. Both Technician A and Technician B
   D. Neither Technician A nor Technician B

8. Technician A says that tower collars allow you to adjust the height of the traction directions. Technician B says that the collar will establish the height of the pulling. Who is correct?
   A. Technician A
   B. Technician B
   C. Both Technician A and Technician B
   D. Neither Technician A nor Technician B

9. Technician A says that nylon straps are used in some pulls because they are as strong as chains. Technician B says that nylon straps are used to protect the parts from damage. Who is correct?
   A. Technician A
   B. Technician B
   C. Both Technician A and Technician B
   D. Neither Technician A nor Technician B

10. Technician A says that the restraint bar is used to maintain a dimension when pulling. Technician B says that a restraint bar is locked into a position to prevent moving. Who is correct?
   A. Technician A
   B. Technician B
   C. Both Technician A and Technician
   D. Neither Technician A nor Technician B

Welded Panel Replacement

Name _____ Date _____ Instructor Review _____

Welded Panel Identification

Fill in the proper name for the commonly welded-on panels shown in the figure that follows.

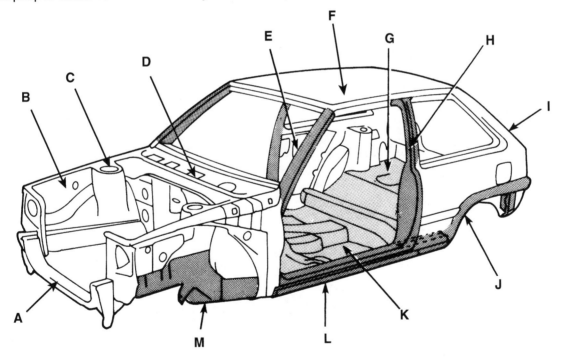

A. _____

B. _____

C. _____

D. _____

E. _____

F. _____

G. _____

H. _____

I. _____

J. _____

K. _____

L. _____

M. _____

Name _____ Date _____

Welding a Rocker Panel

Objective

Upon completion of this activity sheet, the student should be able to safely and accurately splice and weld a rocker panel.

ASE Education Foundation Task Correlation

I.A.4	Select and use a NIOSH-approved air purifying respirator. Inspect condition and ensure fit and operation. Perform proper maintenance in accordance with OSHA regulation 1910.134 and applicable state and local regulations. **(HP-I)**
I.B.8	Remove and replace damaged structural components. **(HP-G)**
I.B.9	Replace protective coatings; restore corrosion protection to repaired or replaced frame areas and anchoring locations. **(HP-G)**
I.B.12	Identify heat limitations and monitoring procedures for structural components. **(HP-G)**
I.C.9	Straighten and align rocker panels and pillars. **(HP-G)**
I.C.10	Straighten and align vehicle openings and floor pans. **(HP-G)**
I.C.13	Identify substrate and repair or replacement recommendations. **(HP-I)**
I.C.16	Determine sectioning procedures of a steel body structure. **(HP-I)**
I.C.17	Remove and replace damaged structural components. **(HP-G)**
I.C.18	Restore corrosion protection to repaired or replaced structural areas and anchoring locations. **(HP-I)**
I.C.19	Determine the extent of damage to aluminum structural components; repair, weld, or replace. **(HP-G)**
I.C.20	Analyze and identify crush/collapse zones. **(HP-I)**
II.C.2	Inspect, remove, and replace mechanically fastened welded steel panel or panel assemblies. **(HP-G)**
VI.A.1	Select and use proper personal safety equipment; take necessary precautions with hazardous operations and materials in accordance with federal, state, and local regulations. **(HP-I)**
VI.A.2	Locate procedures and precautions that may apply to the vehicle being repaired. **(HP-I)**
VI.A.3	Identify vehicle system hazard types (supplemental restraint system [SRS], hybrid/electric/alternative fuel vehicles), locations, and recommended procedures before inspecting or replacing components. **(HP-I)**
VI.A.4	Select and use a NIOSH-approved air purifying respirator. Inspect condition and ensure fit and operation. Perform proper maintenance in accordance with OSHA regulation 1910.134 and applicable state and local regulations. **(HP-I)**
VI.B.1	Identify the considerations for cutting, removing, and welding various types of steel, aluminum, and other metals. **(HP-G)**
VI.B.2	Determine the correct GMAW welder type, electrode/wire type, diameter, and gas to be used in a specific welding situation. **(HP-I)**
VI.B.3	Set up, attach work clamp (ground), and adjust the GMAW welder to "tune" for proper electrode stickout, voltage, polarity, flow rate, and wire-feed speed required for the substrate being welded. **(HP-I)**

VI.B.5   Determine the proper angle of the gun to the joint and direction of gun travel for the type of weld being made. **(HP-G)**

VI.B.6   Protect adjacent panels, glass, vehicle interior, etc. from welding and cutting operations. **(HP-I)**

VI.B.7   Identify hazards; foam coatings and flammable materials prior to welding/cutting procedures. **(HP-G)**

VI.B.8   Protect computers and other electronics/wires during welding procedures. **(HP-I)**

VI.B.9   Clean and prepare the metal to be welded, assure good metal fit-up, apply weld-through primer if necessary, clamp or tack as required. **(HP-I)**

VI.B.10   Determine the joint type (butt weld with backing, lap, etc.) for weld being made. **(HP-I)**

VI.B.11   Determine the type of weld (continuous, stitch weld, plug, etc.) for each specific welding operation. **(HP-I)**

VI.B.12   Perform the following welds: plug, butt weld with and without backing, and fillet in the flat, horizontal, vertical, and overhead positions. **(HP-I)**

VI.B.13   Perform visual evaluation and destructive test on each weld type. **(HP-I)**

VI.B.14   Identify the causes of various welding defects; make necessary adjustments. **(HP-I)**

VI.B.15   Identify cause of contact tip burnback and failure of wire to feed; make necessary adjustments. **(HP-I)**

VI.B.16   Identify cutting process for different substrates and locations; perform cutting operation. **(HP-I)**

We Support

 Education Foundation

Tools and Equipment

18-inch section of salvage rocker panel
Reciprocating saw or air saw
Cutoff tool
Spot weld cutter
Weld-through primer
MIG welder
Appropriate drill bit
Abrasive wheel

Safety Equipment

Safety glasses or goggles
Welding helmet with #10 shade
Welding gloves
Welding jacket or sleeves
Welding respirator

Introduction

Welding is a crucial part of the collision repair process. It is critical that all surfaces that are to be welded be properly researched, planned, and prepared correctly. The structural integrity and the safe operation of the vehicle depend on the quality of the welds. MIG welding is the accepted standard in the automotive repair industry for most mild steels, with new technology and equipment coming into the industry every day. High-strength steels require a different wire type and gas to be welded safely and securely without generating too much heat. Aluminum is making a strong presence in the industry in the past few years as well as more and more manufacturers are making their automobile body panels out of aluminum to reduce the weight of the vehicle so that it cuts down on fuel costs.

Vehicle Description

Year _____ Make _____ Model _____

VIN _____

Procedure

**Task
Completed**

1. Measure in 6 inches from each end of the rocker panel. Mark at the 6-inch lines. Cut at the marks with a cutoff tool or reciprocating/air saw. This will give you a center piece and two ends. ☐

2. The inserts should have the corners trimmed to fit. Remove the paint with an abrasive wheel. Spray on a weld-through primer. ☐

   a. What is the importance of using a weld-through primer? _____

   _____

3. Take notes where the locations of the spot welds are and how many are in each location. Take pictures for proof so when you replace the new part, they go in the exact same locations and the exact same number of spot welds also. ☐

   a. Why is it important that you replace the same number of and locations of spot welds?

   _____

   _____

4. Now drill out the spot welds from the center piece. Be careful to cut only the outer piece. Do not cut a hole through both pieces. When drilling, do not apply excessive pressure. This can be a time-consuming process, but it is the best way to remove a welded panel. You should now have the center piece separated into outer and inner pieces. ☐

   a. Did you go all the way through both panels on any of the spot welds? _____

   b. If so, how will you repair this? _____

   _____

5. Mark the location for ⁵⁄₁₆-inch plug weld holes in the two end pieces. The holes should be ¾ inch away from the edge and put back in the same locations as the previous spot welds were removed. Carefully drill the holes. Remove the burrs from the backside with an angle grinder with a #80 grit disc. ☐

   a. If the backside is not visible, why is it still important to remove the burrs? _____

   _____

6. Test fit the inserts in the two end pieces. ☐

   a. Are there any problems with the fit? If so, explain. _____

   _____

   b. How will you correct this? _____

   _____

7. When they fit correctly, clamp them tightly. ☐

   a. What is the importance of clamping the pieces before welding? _____

   _____

8. Before starting the actual welds, always perform a few test welds to make sure the weld will be sufficient. To do this you will need to make a few 4 inch × 6 inch coupons or test pieces of metal to practice the weld types you are about to perform. Either use a drill and a ⁵⁄₁₆-inch drill bit or a ⁵⁄₁₆-inch hole punch to make all the necessary holes around the edges of one test piece of metal. The other should not have holes in it. Lay the piece with holes down onto the piece without holes. Then clamp them together to fit properly so that they do not shift. ☐

9. Test and tune the welder: Make sure the gas on the welder is turned on and full. Make sure the nozzle does not have any blockages to keep the shielding gas from coming out. Make sure your ground clamp is in the proper location and clamped securely. Now test the welder on an old piece of metal so that all the settings are in the proper locations to perform the necessary test welds.   ☐

   a.   What is the heat setting on? _____

   b.   What is your wire speed? _____

   c.   Did your welder perform properly, or did you have any defects in the welds?

   _____

   _____

10. Now that you have the welder set up correctly, it is time to perform your test welds. Weld all the plug holes so that they are fully filled in and as flush as possible to the existing surface. This will cut down on grind time when grinding down the welds to a smooth surface.   ☐

11. Now it is time to perform a destructive test on your test welds. Place the test welds into a vise and clamp it in place. Bend or roll the metal over the vise so that it stretches the metal and the welded area. If your weld does not break, then it was a good structural weld. If it broke or has a hairline crack through it, your weld is not a good structural solid weld.   ☐

   a.   Did your weld split or break when rolled over the vise? _____

   b.   If so, repeat the earlier steps until your weld is structurally sound and does not split.

12. Now you are ready to start preparing the actual plug weld holes. Drill plug weld holes into the two center pieces. They should be made the same way as the holes on the end pieces. Remove the burrs from the backsides as before.   ☐

13. Align the end pieces and the center pieces. The gap between the end piece and the center piece should be the same as the thickness of the panel.   ☐

14. Measure the gap and make sure it is even all around.   ☐

15. As before, clamp together tightly. Weld all the plug holes so that they are fully filled in and as flush as possible to the existing surface. This will cut down on grind time when grinding down the welds to a smooth surface. Skip weld the butt joints.   ☐

   a.   What is skip welding? _____

   _____

16. Now get your instructor to come and check your welds to see if any corrections need to be made.   ☐

17. When no corrections need to be made, you may begin to practice grinding the welds down to a uniform and smooth surface to see if the entire plug weld is filled in.   ☐

INSTRUCTOR'S COMMENTS _____

_____

_____

Name _____ Date _____

Quarter Panel Replacement

Objective

Upon completion of this activity sheet, the student should be able to safely and accurately replace a quarter panel. This activity should be completed by the advanced student under the close supervision of the instructor. This is likely to be a long-term project.

ASE Education Foundation Task Correlation

I.C.11	Straighten and align quarter panels, wheelhouse assemblies, and rear body sections (including rails and suspension/powertrain mounting points). **(HP-G)**
I.C.13	Identify substrate and repair or replacement recommendations. **(HP-I)**
I.C.15	Repair damage using power tools and hand tools to restore proper contours and dimensions.
I.C.16	Determine sectioning procedures of a steel body structure. **(HP-I)**
I.C.17	Remove and replace damaged structural components. **(HP-G)**
I.C.18	Restore corrosion protection to repaired or replaced structural areas and anchoring locations. **(HP-I)**
I.C.19	Determine the extent of damage to aluminum structural components; repair, weld, or replace. **(HP-G)**
I.C.20	Analyze and identify crush/collapse zones. **(HP-I)**
II.C.2	Inspect, remove, and replace mechanically fastened welded steel panel or panel assemblies. **(HP-G)**
II.C.10	Restore corrosion protection during and after the repair. **(HP-I)**
II.C.13	Perform panel bonding and weld bonding. **(HP-G)**
II.C.16	Weld damaged or torn steel body panels; repair broken welds. **(HP-G)**
II.D.2	Locate and repair surface irregularities on a damaged body panel using power tools, hand tools, and weld-on pulling attachments. **(HP-I)**
II.D.3	Demonstrate hammer and dolly techniques. **(HP-I)**
VI.A.1	Select and use proper personal safety equipment; take necessary precautions with hazardous operations and materials in accordance with federal, state, and local regulations. **(HP-I)**
VI.A.2	Locate procedures and precautions that may apply to the vehicle being repaired. **(HP-I)**
VI.A.3	Identify vehicle system hazard types (supplemental restraint system [SRS], hybrid/electric/alternative fuel vehicles), locations, and recommended procedures before inspecting or replacing components. **(HP-I)**
VI.A.4	Select and use a NIOSH-approved air purifying respirator. Inspect condition and ensure fit and operation. Perform proper maintenance in accordance with OSHA regulation 1910.134 and applicable state and local regulations. **(HP-I)**
VI.B.1	Identify the considerations for cutting, removing, and welding various types of steel, aluminum, and other metals. **(HP-G)**
VI.B.2	Determine the correct GMAW welder type, electrode/wire type, diameter, and gas to be used in a specific welding situation. **(HP-I)**

VI.B.3 Set up, attach work clamp (ground), and adjust the GMAW welder to "tune" for proper electrode stickout, voltage, polarity, flow rate, and wire-feed speed required for the substrate being welded. **(HP-I)**

VI.B.5 Determine the proper angle of the gun to the joint and direction of gun travel for the type of weld being made. **(HP-G)**

VI.B.6 Protect adjacent panels, glass, vehicle interior, etc. from welding and cutting operations. **(HP-I)**

VI.B.7 Identify hazards; foam coatings and flammable materials prior to welding/cutting procedures. **(HP-G)**

VI.B.8 Protect computers and other electronics/wires during welding procedures. **(HP-I)**

VI.B.9 Clean and prepare the metal to be welded, assure good metal fit-up, apply weld-through primer if necessary, clamp or tack as required. **(HP-I)**

VI.B.10 Determine the joint type (butt weld with backing, lap, etc.) for weld being made. **(HP-I)**

VI.B.11 Determine the type of weld (continuous, stitch weld, plug, etc.) for each specific welding operation. **(HP-I)**

VI.B.12 Perform the following welds: plug, butt weld with and without backing, and fillet etc., in the flat, horizontal, vertical, and overhead positions. **(HP-I)**

VI.B.13 Perform visual evaluation and destructive test on each weld type. **(HP-I)**

VI.B.14 Identify the causes of various welding defects; make necessary adjustments. **(HP-I)**

VI.B.15 Identify cause of contact tip burn-back and failure of wire to feed; make necessary adjustments. **(HP-I)**

VI.B.16 Identify cutting process for different substrates and locations; perform cutting operation. **(HP-I)**

We Support

Education Foundation

Tools and Equipment

Auto with damaged quarter panel
MIG welder
Grinder
Cutoff tool
Spot weld cutter or belt sander

Safety Equipment

Safety glasses or goggles
Dust respirator
Work gloves
Welding helmet with #10 shade
Welding gloves
Welding jacket or sleeves
Welding respirator

Introduction

For many years the quarter panels on a vehicle were welded on at the factory. Some late model vehicles have glued-on quarter panels; some have riveted panels. This activity sheet covers only welded-on panels. Precise measuring and cutting are a priority because quarter panels are costly and time-consuming to replace.

Vehicle Description

Year _____ Make _____ Model _____

VIN _____

Procedure

Task Completed

1. Remove the rear bumper, taillight, and trim. Remove the rubber trim on the lip of the trunk. Locate all the spot welds. Take notes of how many total there are, and take pictures of the locations of them so you know how many and where to put them back in when welding back in place. ☐

2. Make sure the panel is clean, so you do not miss any spot welds. Drill a ⅛-inch pilot hole in the spot welds. Cut the welds out with a spot weld cutter. Make sure you only drill through the outer panel. You may use a belt sander also and will save time. ☐

   a. What is the importance of not cutting through both panels in the spot weld?

   _____

   _____

3. Now the importance of researching the OEM repair procedures is an absolute must before you start any replacement procedures. Make sure to use a Google search for the vehicle you are working on. The OEM repair procedures will list the recommended splice locations and the types of joints. ☐

NOTE: Remember, these steps listed next are general steps and are in no way the **EXACT** steps to take on the vehicle you are working on.

4. Decide on the cut location and scribe a rough line cut. ☐

5. Rough cut the damaged quarter panel with a cutoff tool. Stay at least 2 inches away from the rough cut line. At this point the majority of the panel should be cut away. ☐

   a. Why is it important to stay away from the rough cut line? _____

   _____

   b. Is there any other damage that could not be seen with the panel on? _____

   c. If so, describe the damage. _____

   _____

6. Hammer and dolly any damage to the proper contour. The cut metal on the panel is razor sharp, so be very careful. Examine the replacement quarter panel for dents. ☐

7. Scribe a final cut line in the new quarter panel. Double-check your measurements. Cut at the line with your cutoff tool. ☐

   a. How did you hold down the panel when using the cutoff tool?

8. This is a critical step. Once the damaged quarter panel is cut to fit the replacement, any mistakes in measuring will be difficult to correct. ☐

   a. How can measuring mistakes be corrected? _____

   _____

9. Put the replacement quarter panel on the car. Check it for fit and align as necessary. ☐

10. When you are satisfied with the fit, scribe a line on the damaged quarter panel, using the final cut line edge on the replacement quarter panel as a guide. Measure ½ inch in from this scribed line. This gives a ½-inch overlap for splicing. This is the final cut line. Cut on this line with a cutoff tool and grind off the paint. ☐

    a. Why is the ½-inch overlap so important? _____

    _____

11. Apply weld-through primer. Punch plug weld holes, using the same number and location as the factory spot welds so that you do not change the way the vehicle is designed to collapse upon impact. ☐

Task Completed

12. Fit up the replacement panel. Check all the gaps. Realign as necessary.   ☐

13. When the fit is perfect, vise grips or clamps can be used to hold the metal together in other areas. If clamps cannot reach, use sheet metal screws.   ☐

14. Tune and test the welder for appropriate settings to weld the new panel in place.   ☐

15. Weld on the replacement panel. Start with the plug welds. Do not overheat the panel. On the spliced joint, skip weld allowing the surface to cool in between each weld until completely welded.   ☐

    a.   What will happen to the panel if overheated? _____

       _____

    b.   What is the purpose of skip welding? _____

       _____

16. Dress the welds by grinding them down with a grinder. Grind only in the weld, not the surrounding metal.   ☐

17. Remove all the weld scale and blistered paint on the inside of the spliced joint. Scuff the inside of the replacement panel so that it may be sprayed with epoxy primer or self-etching primer depending on the manufacturer's recommendations.   ☐

18. Refer to the priming job sheet for the priming procedures. Once primed, refer to the seam sealer job sheet to seal off any necessary areas.   ☐

INSTRUCTOR'S COMMENTS _____

_____

_____

Review Questions

Name _____ Date _____ Instructor Review _____

1. _____ panels are manufactured by smaller companies and are not OEM.

2. When locating spot welds, Technician A says it is best to use an oxyacetylene torch to remove the paint. Technician B says this could warp the metal. Who is correct?
   A. Technician A
   B. Technician B
   C. Both Technician A and Technician B
   D. Neither Technician A nor Technician B

3. _____ are used to add rigidity to structural parts.

4. Certain structural components have crush zones.
   A. True
   B. False

5. Full body sectioning involves replacing a quarter panel.
   A. True
   B. False

6. Technician A says that welded panels must be MIG welded only. Technician B says that welded panels must be welded with silicon bronze only. Who is correct?
   A. Technician A
   B. Technician B
   C. Both Technician A and Technician B
   D. Neither Technician A nor Technician B

7. Technician A says that a belt sander can be used to remove spot welds. Technician B says that a spot weld remover bit can be used. Who is correct?
   A. Technician A
   B. Technician B
   C. Both Technician A and Technician B
   D. Neither Technician A nor Technician B

8. Technician A says that all late model vehicles' quarter panels are welded on. Technician B says that all late model vehicles' quarter panels are held with two-part epoxy. Who is correct?
   A. Technician A
   B. Technician B
   C. Both Technician A and Technician B
   D. Neither Technician A nor Technician B

9. Technician A says that sectioning involves cutting the part in a location other than the factory seam. Technician B says that sectioning involves cutting the part on the seam. Who is correct?
   A. Technician A
   B. Technician B
   C. Both Technician A and Technician B
   D. Neither Technician A nor Technician B

10. Technician A says OEM repair information is available for all major structural panels. Technician B says OEM repair procedures often give recommendations about how to install frame rail replacements. Who is correct?
    A. Technician A
    B. Technician B
    C. Both Technician A and Technician B
    D. Neither Technician A nor Technician B

Restoring Corrosion Protection

Name _____ Date _____ Instructor Review _____

Corrosion Material Identification

Fill in the proper name for each layer of material shown in the figure that follows.

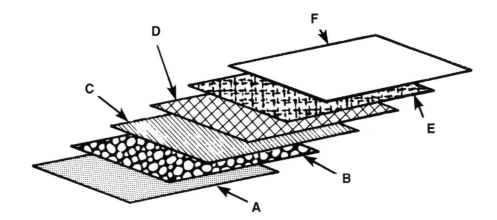

A. _____

B. _____

C. _____

D. _____

E. _____

F. _____

Name _____   Date _____

Surface Rust Removal

Objective
Upon completion of this activity sheet, the student should be able to recognize and repair rust damage.

ASE Education Foundation Task Correlation

II.B.7	Soap and water wash entire vehicle; complete pre-repair inspection checklist. **(HP-I)**
II.B.9	Remove corrosion protection, undercoating, sealers, and other protective coatings as necessary to perform repairs. **(HP-I)**
II.C.10	Restore corrosion protection during and after the repair. **(HP-I)**
II.D.1	Prepare a panel for body filler by abrading or removing the coatings; featheredge and refine scratches before the application of body filler. **(HP-I)**
IV.A.1	Select and use proper personal safety equipment; take necessary precautions with hazardous operations and materials according to federal, state, and local regulations. **(HP-I)**
IV.A.2	Identify safety and personal health hazards according to OSHA guidelines and the "Right to Know Law." **(HP-I)**
IV.A.3	Inspect spray environment and equipment to ensure compliance with federal, state, and local regulations and for safety and cleanliness hazards. **(HP-I)**
IV.A.4	Select and use a NIOSH-approved air purifying respirator. Inspect condition and ensure fit and operation. Perform proper maintenance in accordance with OSHA Regulation 1910.134 and applicable state and local regulations. **(HP-I)**
IV.A.5	Select and use a NIOSH-approved supplied air (Fresh Air Make-up) respirator system. Perform proper maintenance in accordance with OSHA Regulation 1910.134 and applicable state and local regulations. **(HP-I)**
IV.A.6	Select and use the proper personal safety equipment for surface preparation, spray gun and related equipment operation, paint mixing, matching and application, paint defects, and detailing (gloves, suits, hoods, eye and ear protection, etc.). **(HP-I)**
IV.B.2	Soap and water wash entire vehicle; use appropriate cleaner to remove contaminants. **(HP-I)**
IV.B.4	Remove paint finish as needed. **(HP-I)**
IV.B.7	Apply suitable metal treatment or primer in accordance with total product systems. **(HP-I)**
IV.B.8	Mask and protect other areas that will not be refinished. **(HP-I)**
IV.B.9	Demonstrate different masking techniques (recess/back masking, foam door type, etc.). **(HP-G)**
IV.B.10	Mix primer, primer-surfacer, and primer-sealer. **(HP-I)**
IV.B.12	Apply primer onto surface of repaired area. **(HP-I)**
IV.B.16	Remove dust from area to be refinished, including cracks or moldings of adjacent areas. **(HP-I)**
IV.B.17	Clean area to be refinished using a final cleaning solution. **(HP-I)**
IV.B.18	Remove, with a tack rag, any dust or lint particles from the area to be refinished. **(HP-I)**

We Support
ASE | **Education Foundation**

Tools and Equipment

Rusted vehicle
#80, #180, #320 grit sandpaper
Grinder
Scratch awl or pick tool
Metal prep
Sandblaster (if applicable)
Coarse scratch pad
Self-etching primer
Hard block
Razor blade

Safety Equipment

Safety glasses or goggles
Dust respirator
Work gloves

Introduction

Even though the painting process and materials have improved largely over the years, vehicles still rust. This can be due to scratches or anything that breaks the original finish. Most collision repair centers try to avoid doing rust repair because it is hard to guarantee. However, there are some collision centers that specialize in rust repair.

Vehicle Description

Year _____ Make _____ Model _____

VIN _____

Procedure

Task Completed

1. Wash the vehicle and examine it for rust. Rust usually begins in the door hem flanges where the body panels meet or under the moldings. Occasionally, it occurs in the center of the panel due to paint damage. Bubbles in the paint indicate rust. ☐

   a. What are the panels and areas of rust on the vehicle? Be sure to note whether there are rust holes or surface rust. _____

   _____

2. Probe any bubbles with a scratch awl or pick. You will see dark-colored steel in them; this is rust. Many times, the metal is so weak that the awl or pick will pierce the metal. ☐

   a. If you go through the metal, how will you repair the hole? _____

   _____

3. If the awl does not pierce the metal, you have found surface rust. Use the grinder to remove the paint from the area. Sometimes the rust will extend outward from a central location. Make sure you find all traces of rust. Featheredge the paint edge with #80 and #180 grit sandpaper. ☐

   a. When you have finished with the grinder, how do you know you have removed all the rust? _____

   _____

**Task
Completed**

4. There will be numerous pits in the metal. Use a wire wheel to remove all the sur-   ☐
face rust that was not removed in the grinding process. If you are going to use a
sandblaster, blast until the dark spots have been removed from the pits.

5. Apply metal etch with a spray bottle. Wear rubber gloves and safety glasses.   ☐
Work the area with a coarse scuff pad. Keep the surface wet for five minutes.
Wipe off the area. Wipe again with a damp cloth and wipe dry. Or you can use a
rust converter to convert the metal to a black primer.

6. Mask off the area. Mix the self-etching primer. Spray two wet coats on the bare
metal. Allow proper flash time between coats.

    a. What is the purpose of self-etching primer? _____

_____

_____

NOTE: After the surface rust has been removed, it will need to be primed with a
urethane 2k primer and any surface imperfections filled in with the appropriate body
filler or spot putty. These procedures are covered in another job sheet.

INSTRUCTOR'S COMMENTS _____

_____

_____

Name _____ Date _____

Corrosion Prevention—Enclosed Rail

Objective
Upon completion of this activity sheet, the student should be able to properly treat a spliced rail or corroded metal panel for prevention of corrosion.

ASE Education Foundation Task Correlation

I.A.1   Select and use proper personal safety equipment; take necessary precautions with hazardous operations and materials in accordance with federal, state, and local regulations. **(HP-I)**

I.A.2   Locate procedures and precautions that may apply to the vehicle being repaired. **(HP-I)**

I.A.4   Select and use a NIOSH-approved air purifying respirator. Inspect condition and ensure fit and operation. Perform proper maintenance in accordance with OSHA regulation 1910.134 and applicable state and local regulations. **(HP-I)**

I.B.9   Replace protective coatings; restore corrosion protection to repaired or replaced frame areas and anchoring locations. **(HP-G)**

I.C.18   Restore corrosion protection to repaired or replaced structural areas and anchoring locations. **(HP-I)**

II.C.10   Restore corrosion protection during and after the repair. **(HP-I)**

IV.A.1   Select and use proper personal safety equipment; take necessary precautions with hazardous operations and materials according to federal, state, and local regulations. **(HP-I)**

IV.A.2   Identify safety and personal health hazards according to OSHA guidelines and the "Right to Know Law." **(HP-I)**

IV.A.3   Inspect spray environment and equipment to ensure compliance with federal, state, and local regulations and for safety and cleanliness hazards. **(HP-I)**

IV.A.4   Select and use a NIOSH-approved air purifying respirator. Inspect condition and ensure fit and operation. Perform proper maintenance in accordance with OSHA Regulation 1910.134 and applicable state and local regulations. **(HP-I)**

IV.A.5   Select and use a NIOSH-approved supplied air (Fresh Air Make-up) respirator system. Perform proper maintenance in accordance with OSHA regulation 1910.134 and applicable state and local regulations. **(HP-I)**

IV.A.6   Select and use the proper personal safety equipment for surface preparation, spray gun and related equipment operation, paint mixing, matching and application, paint defects, and detailing (gloves, suits, hoods, eye and ear protection, etc.). **(HP-I)**

We Support
ASE | Education Foundation

Tools and Equipment
Spliced frame rail or corroded metal panel
#180 grit paper
Rustproofing wax
Rustproofing 360 spray wand
Blowgun
Abrasive pad

Safety Equipment
Safety glasses or goggles
NIOSH-approved paint respirator
Impervious gloves

Introduction

Vast improvements have been made in the auto industry to slow down corrosion, but weather conditions around the world still cause vehicles to rust. If the proper steps and materials are used, rust repair can last indefinitely. Make sure to use all PPE for this exercise.

Vehicle Description

Year _____ Make _____ Model _____

VIN _____ or Type of panel and area being treated _____

Procedure

Task Completed

1. This exercise involves a spliced joint that is not accessible whether it be an enclosed frame rail or corroded metal panel. Accessible joints—those visible at the front of the rail—are treated in the same manner.

2. Remove the weld scale from the joint by directing the air from a blowgun into the rail. You may also use a long-handled brush to reach into the rail to remove the heat-blistered paint.   ☐

   a. What can happen if you do not remove the weld scaling? _____
   _____

3. Clean the outside of the rail at the weld joint with an abrasive pad. Remove all the blistered paint. Featheredge the paint with #180 grit sandpaper.   ☐

4. Shake the aerosol can of corrosion protection for at least one minute to properly mix the components inside. Attach the 360 wand and extensions to the desired length needed to reach the inner repair locations.   ☐

   a. What can happen if you do not shake the can before applying? _____
   _____

5. Insert the wand of the corrosion protection into the rail. Make sure the wand is inserted beyond the joint. Pull the trigger to start the spray. Pull the wand toward you. Release the trigger before you reach the end of the rail. Spray up to three coats to ensure full coverage and maximize protection.   ☐

   a. Why is it important to spray beyond the joint if it is enclosed? _____
   _____

   b. Why is it necessary to spray up to three coats? _____
   _____

6. Clean the nozzle after you are done. Put the spray nozzle with the wand attachment onto a can of spray gun cleaner, and spray until the nozzle and wand are clean throughout.   ☐

   a. What will happen to the spray wand and nozzle if you do not clean it out? _____
   _____

7. Remove any excess that may drip or run out of any access holes with an appropriate wax and grease remover.   ☐

8. Spray the outside of the rail with the epoxy primer, using a conventional spray gun. Allow the proper flash time, and then spray the outside of the rail with a topcoat in a conventional spray gun. Allow the paint to dry overnight.   ☐

**Task
Completed**

9. You may use the rustproofing wax or cavity wax in the same way you sprayed the ☐
   rustproofing into the enclosed rail as before.

INSTRUCTOR'S COMMENTS _____

_____

_____

Name _____ Date _____

Seam Sealer Removal and Application

Objective

Upon completion of this activity sheet, the student should be able to restore seam sealer back to the repaired areas of the damaged automobile or test panel. Seam sealer is a crucial part of the corrosion protection process in today's automobiles. It is a key to keep moisture out of the two layers of metal that are joined together and to also keep any harmful fumes from entering the passenger compartment from outside the vehicle.

ASE Education Foundation Task Correlation

I.C.18   Restore corrosion protection to repaired or replaced structural areas and anchoring locations. **(HP-I)**

II.B.9   Remove corrosion protection, undercoating, sealers, and other protective coatings as necessary to perform repairs. **(HP-I)**

IV.B.22   Restore caulking and seam sealers to repaired areas. **(HP-G)**

We Support

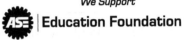 | Education Foundation

Tools and Equipment

Metal workbench to practice
Clear packing tape
Automobile or test panel with area of
    seam sealer exposed
Wire wheel in an eraser wheel or a drill
Coarse scotch pad or #320 grit sandpaper
Self-etching primer
Seam sealer
Seam sealer dispenser
Masking tape
Body filler spreaders

Safety Equipment

Safety glasses or goggles
Dust respirator
Work gloves
Rubber gloves

Introduction

Even though the painting process and materials have improved greatly over the years, vehicles still rust. This can be due to scratches or anything that breaks the original finish, especially through the seam sealer. This job sheet will help you duplicate the original OEM seam sealer that was applied from the factory.

Vehicle Description

Year _____ Make _____ Model _____

VIN _____ or Test panel _____

Procedure

1. It is difficult to explain how to match the appearance of OEM seam sealer if you have never seen it. From here you will need to go to YouTube to watch a few videos to get familiar with different applications and techniques.

2. Locate the YouTube video titled "Seam Sealer Matching Made Easy" by the 3M Collision Repair Channel.

    a. Note the different techniques they show you. Find a metal workbench, and line a section with the clear packing tape like the video. Tape off sections on the clear packing tape, and try each of the different applications shown in the video. These are all easy to replicate and very important to learn how to do in the collision industry.   ☐

    b. Did your seam sealer applications match the video appearances? _____

    c. If not, explain why and how you will correct them to make them match properly.

    _____

    _____

    _____

    d. Once completed, have your instructor check your work.

3. Now, recreate the ones that did not match properly, using your preceding statement and your instructor's input.   ☐

    e. Did they match correctly this time? Explain why or why not.

    _____

    _____

    _____

4. There are several more videos on the 3M Collision Repair Channel you can watch to better understand the procedure.

5. If you have a panel with seam sealer on it, first you must determine which method will best match the OEM seam sealer appearance. Then, get a wire wheel in a drill or an eraser wheel with a wire wheel attachment for fast removal of the old seam sealer.   ☐

    a. Which method did you use to replicate the OEM appearance?

    _____

    _____

6. Tape off the desired outline for the new seam sealer, and prepare the surface for the new seam sealer.   ☐

7. Apply the seam sealer and shape to the desired result to match.   ☐

8. Once completed, get your instructor to check your work.   ☐

INSTRUCTOR'S COMMENTS _____

_____

_____

Review Questions

Name _____ Date _____ Instructor Review _____

1. _____ is the oxidation and chemical change of metal.

2. The process of coating metal with zinc is called _____.

3. _____ is the term given to rain containing pollutants.

4. Heavy road salts accelerate the corrosion process.
   A. True
   B. False

5. Acid rain repairs cannot be made.
   A. True
   B. False

6. Anticorrosion materials are used to prevent rusting of metal parts.
   A. True
   B. False

7. Technician A says that you can drill a hole in a panel to gain access for corrosion protection procedures. Technician B says that you should use a flexible spray wand where access is limited. Who is correct?
   A. Technician A
   B. Technician B
   C. Both Technician A and Technician B
   D. Neither Technician A nor Technician B

8. Technician A says that self-etching primer is sprayed in one very heavy coat. Technician B says that two thin wet coats are sprayed. Who is correct?
   A. Technician A
   B. Technician B
   C. Both Technician A and Technician B
   D. Neither Technician A nor Technician B

9. Technician A says that thin-bodied sealers are used to fill seams ¼ inch or larger. Technician B says that the sealers are used for ½ inch or under. Who is correct?
   A. Technician A
   B. Technician B
   C. Both Technician A and Technician B
   D. Neither Technician A nor Technician B

10. Technician A says that self-etching primers can be used for exposed interior parts. Technician B says that lacquer-based primer will not provide proper adhesion on bare metal. Who is correct?
    A. Technician A
    B. Technician B
    C. Both Technician A and Technician B
    D. Neither Technician A nor Technician B

Chassis Service and Wheel Alignment

Name _____ Date _____ Instructor Review _____

Identifying Basic Parts

Using the following figure, name the parts that correspond to the numbers.

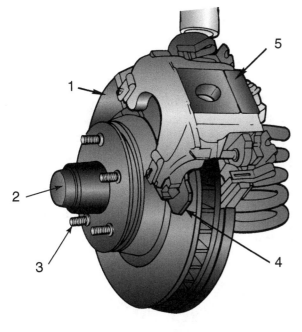

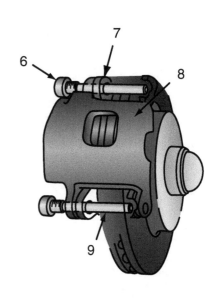

1. _____

2. _____

3. _____

4. _____

5. _____

6. _____

7. _____

8. _____

9. _____

Name _____ Date _____ Instructor Review _____

Identifying Basic Suspension Parts

Using the following figure, name the parts that correspond to the numbers.

A

1. _____	5. _____
2. _____	6. _____
3. _____	7. _____
4. _____	

B

1. _____	4. _____
2. _____	5. _____
3. _____	6. _____

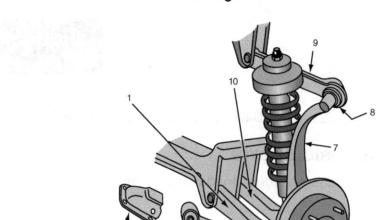

C

1. _____   6. _____

2. _____   7. _____

3. _____   8. _____

4. _____   9. _____

5. _____   10. _____

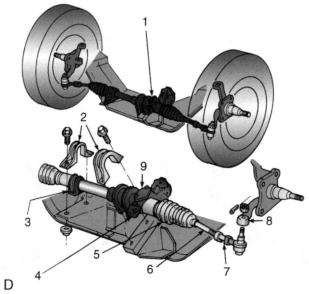

D

1. _____   6. _____

2. _____   7. _____

3. _____   8. _____

4. _____   9. _____

5. _____

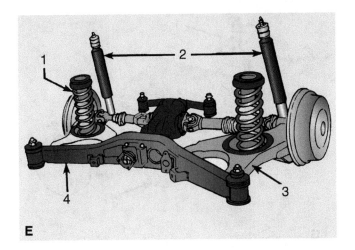

1. _____   3. _____

2. _____   4. _____

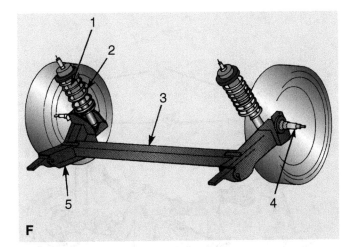

1. _____   4. _____

2. _____   5. _____

3. _____

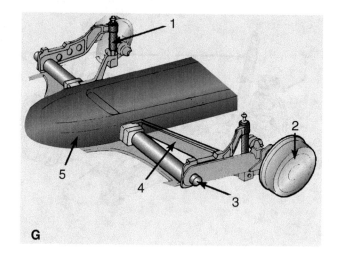

G

1. _____   4. _____

2. _____   5. _____

3. _____

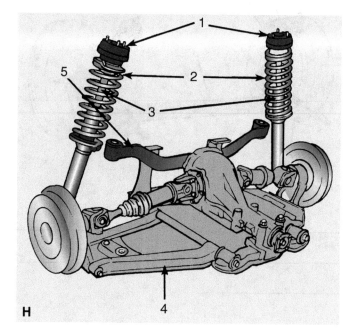

H

1. _____   4. _____

2. _____   5. _____

3. _____

Name _____ Date _____ Instructor Review _____

Identifying Basic Cooling System Parts

Using the figures below, name the parts that correspond to the numbers.

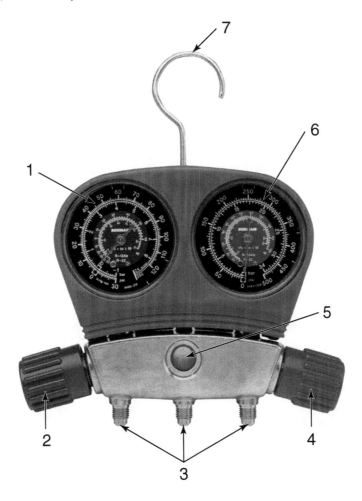

1. _____

2. _____

3. _____

4. _____

5. _____

6. _____

7. _____

Name _____ Date _____

Suspension and Steering Checks

Objective

Upon completion of this activity sheet, the student should be able to accurately diagnose steering noise in power steering, rack and pinion, and MacPherson struts. The student should also be able to explain the needed repairs.

ASE Education Foundation Task Correlation

III.A.1	Select and use proper personal safety equipment; take necessary precautions with hazardous operations and materials in accordance with federal, state, and local regulations. **(HP-I)**
III.A.2	Locate procedures and precautions that may apply to the vehicle being repaired. **(HP-I)**
III.A.3	Identify vehicle system hazard types (supplemental restraint system [SRS], hybrid/electric/alternative fuel vehicles), locations, and recommended procedures before inspecting or replacing components. **(HP-I)**
III.A.4	Select and use a NIOSH-approved air purifying respirator. Inspect condition and ensure fit and operation. Perform proper maintenance in accordance with OSHA regulation 1910.134 and applicable state and local regulations. **(HP-I)**
III.B.1	Perform visual inspection and measuring checks to identify steering and suspension collision damage. **(HP-G)**
III.B.2	Identify one-time-use fasteners. **(HP-I)**
III.B.3	Clean, inspect, and prepare reusable fasteners. **(HP-I)**
III.B.19	Measure vehicle ride height and wheel base; determine necessary action. **(HP-I)**
III.B.22	Verify proper operation of steering systems including electronically controlled, hydraulic, and electronically assisted steering systems. **(HP-G)**
III.B.23	Diagnose front and rear suspension system noises and body sway problems; determine necessary action. **(HP-G)**
III.B.24	Diagnose vehicle wandering, pulling, hard steering, bump steer, memory steering, torque steering, and steering return problems; determine necessary action. **(HP-G)**
III.B.26	Diagnose tire wear patterns; determine cause. **(HP-I)**
III.B.27	Inspect tires; identify direction of rotation and location; check tire size and tire pressure monitoring system (TPM) and adjust air pressure. **(HP-I)**
III.B.28	Diagnose wheel/tire vibration, shimmy, tire pull (lead), wheel hop problems; determine needed repairs. **(HP-G)**
III.B.30	Reinstall wheels and torque lug nuts. **(HP-I)**

We Support

ASE | Education Foundation

Tools and Equipment

Vehicle
Appropriate service manual
Assorted hand tools
Tape measure
½-inch impact wrench
Appropriate impact sockets

Safety Equipment

Safety glasses with side shields or goggles

Introduction

Steering and suspension are critical components in the safe operation of a vehicle. A vehicle has many moving parts in the steering and suspension that must operate safely for years. This activity sheet is designed for the advanced student and should be done under the instructor's supervision.

Vehicle Description

Year _____ Make _____ Model _____

VIN _____

Procedure

Task Completed

1. Check the power steering fluid level. ☐

   a. What is the appearance of the fluid? _____

2. Check wheel lugs for looseness. ☐

   a. How is this done? _____
   _____

3. Check the tire size and the air pressure in all four tires to make sure they are all to the proper specifications. ☐

   a. Was the air pressure correct in all four tires? _____

   b. What was the recommended air pressure for the tires? _____

4. Remove a wheel and inspect the lug nuts and bolt threads for any bent, stripped, or damaged bolts or threads. ☐

   a. If damaged, how will you correct the damage? _____
   _____

5. Reinstall the wheel and torque the lug nuts to the proper specification. ☐

   a. What were the torque specifications? _____

6. Identify some of the one-time-use fasteners used in suspension components. _____ ☐
   _____

7. Measure the vehicle ride height and wheel base. Measure from the center of each wheel well to the middle of each wheel. Write down your measurements for each wheel, and compare the right side measurements to the left side measurements. ☐

   a. What was the left front wheel measurement? _____

   b. What was the right front wheel measurement? _____

   c. What was the right rear wheel measurement? _____

   d. What was the left rear wheel measurement? _____

8. Now measure each side wheel base. Measure from the center of the front wheel on the left side to the center of the rear wheel on the left side. Write down your measurements. Repeat the same steps for the right side. ☐

   a. What were the left side measurements? _____

   b. What were the right side measurements? _____

**Task
Completed**

9. Inspect the tires for any improper tire wear patterns.   ☐

    a.   What wear pattern did you find? _____

    b.   What would cause the tires to wear on the outside edges? _____
_____

    c.   What would cause the tires to wear on the inside edges? _____
_____

    d.   What would cause the tire tread to be flat in the middle but not on the
edges? _____
_____

    e.   What would cause the tire tread to be flat on the edges but not in the
middle? _____
_____

10. Look for the tire pressure monitoring system indicator light (TPM sensor) on the   ☐
instrument panel. Turn on the key, and start the vehicle. Once the engine is started,
watch and see if the light stays on or goes off on its own.

    a.   What would cause it to stay on, and how would you correct it?_____
_____

11. Check the steering belt for looseness.   ☐

    a.   What is the condition of the belt? _____

    b.   Does the belt need to be replaced? _____

12. Check all of the power steering hoses.   ☐

    a.   Is there any cracking? _____

    b.   If so, should the hose(s) be replaced? _____

    c.   Is there any swelling in the hose(s)? _____

    d.   What causes swelling in the hose(s)? _____
_____

    e.   Is there any other type of deterioration? _____

    f.   If so, what is it? _____
_____

13. Bleed the power steering system according to the manufacturer's   ☐
recommendations.

    a.   Did you need a partner to help bleed the power steering? _____

    b.   If so, what did your partner do to help you? _____
_____

    c.   Did you need a special tool to bleed the power steering? _____

    d.   If so, what tool did you use? _____

14. With the vehicle running, turn the steering wheel through its full range of motion. ☐
    a.　Is there any binding? _____
    b.　Is there any uneven turning effort? _____
    _____
    c.　If so, describe. _____
    _____
    d.　Is there any hard steering? _____
    e.　If so, describe. _____
    _____

15. Check the vehicle for steering fluid leaks. ☐
    a.　Have you found any leaks? _____
    b.　If so, where is the fluid leaking from? _____
    _____
    c.　If parts need to be replaced, what are they? _____
    _____

16. With your instructor, test drive the vehicle. ☐
    a.　Is there any body sway? _____
    b.　If so, describe. _____
    _____
    c.　Is there any wandering? _____
    d.　If so, describe. _____
    _____
    e.　Is there pulling? _____
    f.　If so, describe. _____
    _____
    g.　Is there any hard steering? _____
    h.　If so, describe. _____
    _____

17. Check for bump steer and torque steer. ☐
    a.　Are there any problems? _____
    b.　How will you repair any problems? _____
    _____

INSTRUCTOR'S COMMENTS _____
_____
_____

Name _____ Date _____

Disc Brake Inspection and Service

Objective

Upon completion of this activity sheet, the student should be able to inspect brake lines, hoses, and fittings. Along with being able to remove and replace disc brakes, calipers, rotor assemblies, and all appropriate relayed hardware, hoses, and fittings, students should also be able to inspect the parking brake system operation and repair or adjust as necessary.

ASE Education Foundation Task Correlation

III.A.1	Select and use proper personal safety equipment; take necessary precautions with hazardous operations and materials in accordance with federal, state, and local regulations. **(HP-I)**
III.A.2	Locate procedures and precautions that may apply to the vehicle being repaired. **(HP-I)**
III.A.3	Identify vehicle system hazard types (supplemental restraint system [SRS], hybrid/electric/alternative fuel vehicles), locations, and recommended procedures before inspecting or replacing components. **(HP-I)**
III.A.4	Select and use a NIOSH-approved air purifying respirator. Inspect condition and ensure fit and operation. Perform proper maintenance in accordance with OSHA regulation 1910.134 and applicable state and local regulations. **(HP-I)**
III.D.1	Inspect brake lines, hoses, and fittings for damage or wear; tighten fittings and supports; and replace brake lines (double flare and ISO types). **(HP-G)**
III.D.2	Replace hoses, fittings, seals, and supports. **(HP-I)**
III.D.3	Identify, handle, store, and fill with appropriate brake fluids. **(HP-G)**
III.D.4	Bleed (manual, pressure, or vacuum) hydraulic brake system. **(HP-I)**
III.D.6	Adjust brake shoes or pads; remove and reinstall brake drums or drum/hub assemblies. **(HP-I)**
III.D.7	Remove, clean, and inspect caliper and rotor assembly and mountings for wear and damage; reinstall. **(HP-I)**
III.B.8	Inspect parking brake system operation, repair or adjust as necessary, and verify operation. **(HP-I)**

We Support

ASE | Education Foundation

Tools and Equipment

Vehicle
Service jack or a vehicle lift
Appropriate hand tools and torque wrench
Disc brake pad spreader set or C-clamps
Brake pads and rotors (if replacing)
Brake fluid
Brake cleaner

Safety Equipment

Safety glasses with side shields or goggles
Appropriate dust mask
Appropriate protective gloves

Introduction

Brakes are one of the most overlooked areas in most people's everyday lives. Just because they are still stopping the car does not mean they are working at optimal performance. This job sheet will show you how to inspect and service disc brake systems.

Vehicle Description

Year _____ Make _____ Model _____

VIN _____

Procedure

Task Completed

1. Park the vehicle on a flat, dry surface, and install any necessary wheel chocks. Be careful when lifting the vehicle. Always use secure jack points and the appropriate weight-rated jack stands for lowering the vehicle. Be sure to wear safety glasses/goggles and protective gloves.   ☐

   a. What can happen if brake dust gets in your eyes or is breathed into your lungs?

   _____

   _____

2. Raise the hood, and check the brake fluid level in the reservoir. If it is full, some fluid will have to be removed. This will help you avoid spilling any fluid when compressing the brake caliper piston.   ☐

   a. How will you safely remove a small amount of fluid? _____

   _____

3. Loosen the lug nuts with a breaker bar and socket while the wheels are on the ground. Raise the vehicle from a secure lifting point, and secure it on the appropriate jack stands. Remove the lug nuts and the wheel.   ☐

4. You should be able to reach the caliper bolts easily. Turning the wheel left or right will provide easier access. Once the caliper is removed, suspend it with a bungee cord, zip tie, or other means to keep it from hanging freely. (Removing the caliper may require different steps on each vehicle.)   ☐

   a. Why is it important to secure the caliper with bungee cords or a zip tie? _____

   _____

5. Inspect the old pads for uneven wear patterns. The brake pad box will have a guide to reference when inspecting for uneven wear.   ☐

6. Once the pads are off, remove the rotor. On most vehicles, you will have to remove the caliper mounting bracket in order to remove the rotor. When removing the rotor, watch out for rust or uneven wear. You may need help from a rust penetrant or a rubber mallet to finish removing it from the hub assembly.   ☐

7. The hub surface needs to be cleaned with a wire brush to remove any rust and/or debris. This will ensure the new rotor sits flush on the hub and eliminates the possibility of pedal pulsation. Rust as thin as a sheet of paper can translate to pedal pulsation.   ☐

8. Clean the new rotor with appropriate brake cleaner and wipe clean. Now install the new rotor.   ☐

   a. Why is it important to clean the new rotor before installing? _____

   _____

**Task
Completed**

9. Replace the old brake hardware with new hardware. When worn, old brake hardware creates vibrations that are heard as a squealing brake noise at low stopping speeds. These clips should be replaced with each brake job. Make sure to apply brake lubricant at the sliders and the slider contact points before installing the hardware. ☐

    a.   What will happen if you do not apply brake lubricant to the sliders and contact points? _____

_____

10. Install the new brake pads. Pay attention to the wear-sensor position placement to install the pads correctly. Different pads have varying wear-sensor placement. Some pad sets have wear sensors on just the inner pads, some on all four pads, and some do not have wear sensors at all. ☐

11. Inspect the brake caliper and piston for any brake fluid seepage. If the caliper and piston are clean, use the disc brake pad spreader set or C-clamps to safely seat the caliper piston into position. On some rear disc brakes, the caliper pistons screw in and require the disc brake pad spreader set to be seated properly. ☐

12. Take the caliper off the bungee cord or zip tie, and put it back on. **DO NOT OVERTIGHTEN** the caliper bolts. This can lead to much worse problems. Start tightening the bolts with a socket, and then use a torque wrench to tighten the bolts correctly. ☐

    a.   What can happen if the bolts are overtightened? _____

_____

13. Top-off the fluid reservoir with the appropriate fluid to the maximum fill line. Be sure the brake fluid reservoir cap is installed before going to the next step. This will avoid spillage. ☐

14. Sit in the vehicle, and slowly pump the brake pedal a few times until it feels firm. This will make sure brake fluid is flowing properly and the system is working. ☐

    a.   Did the pedal build up pressure? If not, what could be the problem?_____

_____

15. Top-off the fluid in the reservoir again if necessary. Always reinstall the cap after topping off to avoid spillage. ☐

Bleeding the Brakes

16. You must bleed the brakes now to release the air that has got into the system. ☐

17. Ask a friend or fellow student to sit at the wheel. Put on your protective eyewear and gloves to protect against accidental contact with brake fluids. ☐

18. Place a bucket or bowl of some type below the bleeder valve. ☐

19. Use a wrench to open the bleeder valve (size of wrench needed varies by manufacturer). ☐

**Task
Completed**

20. As you open the bleeder valve, ask your assistant to press down slowly on the ☐
brake pedal. Some brake fluid will be lost during this process. Do not worry, this is
ok. The escaping air bubbles will pop or hiss as they come out.

    a.   What is the reason for pushing the pedal slowly instead of fast? _____

_____

21. Close the bleeder valve before your helper eases off the brake pedal. Repeat ☐
several times until the brake fluid pours out consistently without spitting air with the
fluid or makes any hissing or bubbling sounds at all.

22. Top-off the brake fluid reservoir to the maximum fill line. Ensure the reservoir does ☐
not get too low and allow air to be pulled back into the system during bleeding.
Always reinstall the cap after topping off to avoid spillage.

23. Repeat these steps for each wheel to ensure all the air has been removed from the ☐
system.

24. Have your instructor check your work before installing the wheel and lug nuts back ☐
on the vehicle.

25. Install the wheel, and tighten the lug nuts while the vehicle is still raised. Next, raise ☐
the vehicle with a jack, remove the jack stands, lower the vehicle until the wheel is
just touching, and tighten the lug nuts with a torque wrench to the proper setting.

INSTRUCTOR'S COMMENTS _____

_____

_____

_____

Name _____ Date _____

Cooling System Service

Objective

Upon completion of this activity sheet, the student should be able to inspect the cooling system heater hoses and belts, radiator pressure cap, coolant recovery system, water pump, and thermostat and refill and test the system.

ASE Education Foundation Task Correlation

III.A.1   Select and use proper personal safety equipment; take necessary precautions with hazardous operations and materials in accordance with federal, state, and local regulations. **(HP-I)**

III.A.2   Locate procedures and precautions that may apply to the vehicle being repaired. **(HP-I)**

III.A.3   Identify vehicle system hazard types (supplemental restraint system [SRS], hybrid/electric/alternative fuel vehicles), locations, and recommended procedures before inspecting or replacing components. **(HP-I)**

III.A.4   Select and use a NIOSH-approved air purifying respirator. Inspect condition and ensure fit and operation. Perform proper maintenance in accordance with OSHA regulation 1910.134 and applicable state and local regulations. **(HP-I)**

III.F.1   Check engine cooling and heater system hoses and belts; determine necessary action. **(HP-I)**

III.F.2   Inspect, test, remove, and replace radiator, pressure cap, coolant system components, and water pump. **(HP-G)**

III.F.3   Recover, refill, and bleed system with proper coolant and check level of protection; leak test system and dispose of materials in accordance with EPA regulations. **(HP-I)**

III.F.4   Remove, inspect, and replace fan (both electrical and mechanical), fan sensors, fan pulley, fan clutch, and fan shroud; check operation. **(HP-G)**

We Support

ASE | Education Foundation

Tools and Equipment

Vehicle
Appropriate service manual
Hand tools
Radiator pressure tester
Hydrometer
Belt tension gauge

Safety Equipment

Safety glasses with side shields or goggles
Protective gloves
Long-sleeved shirt

Introduction

The cooling system on a vehicle is critical to protecting the engine from overheating and being damaged. It is important that coolant be checked and changed on a regular basis. These simple and inexpensive steps can keep a vehicle from breaking down; it can cost thousands of dollars to replace the engine.

Vehicle Description

Year _____ Make _____ Model _____

VIN _____

Procedure	**Task Completed**

1. Drain the coolant from the system, and dispose of the coolant according to OSHA regulations.   ☐

   a. How is the coolant drained? _____

   b. How much coolant does this system hold? _____

2. Check the heater hoses. If there is any damage, hoses should be replaced.   ☐

   a. Are there any cracks, bulges, and/or soft or swollen areas? If so, describe where. _____

_____

3. Check the drive belts.   ☐

   a. Is there any evidence of cracking, tearing, oil soaking, or wear? If so, list here.

_____

   b. What would be the cause of cracking or tearing in a belt? _____

_____

   c. An oil-soaked belt would be an indication of what? _____

_____

4. Using a belt tension gauge, tighten the drive belt to the manufacturer's specification.

   a. Does the vehicle you are working on automatically set the tension? _____

   b. If so, what is the purpose of the gauge? _____

_____

5. Inspect the radiator cap.   ☐

   a. How much pressure can the cap sustain? _____

   b. How is the coolant recovery system inspected? _____

   c. Are there any signs of visible damage? If so, describe. _____

_____

6. Check and inspect the water pump.   ☐

   a. Is there any sign of collision damage? If so, describe. _____

_____

   b. Is there any sign of worn or leaking parts? _____

   c. How do you check the pump for leaking parts? _____

_____

7. Refill the system with the recommended coolant.   ☐

   a. How much coolant was added to the system? _____

   b. What is the purpose of diluting coolant? _____

   c. Will diluting the coolant change the temperature level of protection? Explain.

_____

_____

**Task
Completed**

    d.   How is a hydrometer used to check the level of protection? _____

_____

8. Using a radiator pressure tester, check the system and the cap. ☐

    a.   How is the pressure tester used to check the system? _____

_____

9. Inspect the fan assembly including the fan shroud, fans, and fan clutch for any ☐
    damage. If the engine is able to start and run full of coolant and without damaging
    anything, start the engine and make sure the fans turn on and operate properly.

    a.   If any damage was found, describe it here. _____

_____

INSTRUCTOR'S COMMENTS _____

_____

_____

_____

Name _____ Date _____

A/C System Inspection and Service

Objective

Upon completion of this activity sheet, the student should be able to inspect the air-conditioning (A/C) system and components, as well as recover the oil and refrigerant and recharge the system.

ASE Education Foundation Task Correlation

III.A.1	Select and use proper personal safety equipment; take necessary precautions with hazardous operations and materials in accordance with federal, state, and local regulations. **(HP-I)**
III.A.2	Locate procedures and precautions that may apply to the vehicle being repaired. **(HP-I)**
III.A.3	Identify vehicle system hazard types (supplemental restraint system [SRS], hybrid/electric/alternative fuel vehicles), locations, and recommended procedures before inspecting or replacing components. **(HP-I)**
III.A.4	Select and use a NIOSH-approved air purifying respirator. Inspect condition and ensure fit and operation. Perform proper maintenance in accordance with OSHA regulation 1910.134 and applicable state and local regulations. **(HP-I)**
III.E.3	Locate and identify A/C system service ports. **(HP-I)**
III.E.4	Identify refrigerant contamination; recover, label, and recycle refrigerant from an A/C system. **(HP-G)**
III.E.5	Select refrigerant, evacuate, and recharge A/C system. **(HP-I)**
III.E.6	Select oil type and install correct amount in A/C system. **(HP-I)**
III.E.10	Inspect, test, and replace A/C system condenser and mounts. **(HP-G)**

We Support

ASE | Education Foundation

Tools and Equipment

Vehicle
Appropriate service information
Appropriate hand tools
A/C recovery and recharge machine

Safety Equipment

Safety glasses with side shields or goggles
Protective gloves
Long-sleeved shirt

Introduction

The A/C condenser and lines are very common to be damaged in a collision. They are the very first thing inside the radiator support in most cases, so they are the first to take the impact behind the bumper. This job sheet will show you how to inspect the system for bends, kinks, and other damage, as well as recover the refrigerant and oil from the vehicle, hold a vacuum, and recharge the system.

Vehicle Description

Year _____ Make _____ Model _____

VIN _____

Inspection of the A/C System

Procedure

Task Completed

1. Inspect the A/C condenser. What is the condition of the fins? _____   ☐

   a. Are there any trash or contaminants restricting airflow through the fins? _____

   _____

   b. Are there any kinks or sharp bends in the lines entering and leaving the condenser?

   _____

2. If the condenser fans were clogged with trash, leaves, and debris, how would the   ☐
   A/C system pressures be affected? _____

3. Locate the A/C system service ports.   ☐

   a. Identify which one is the low side and which one is the high side. _____

   _____

Servicing the A/C System

Procedure

4. First, you must locate and identify the type of refrigerant used in this system.   ☐

   a. What is the type of refrigerant? _____

5. Record the level of oil located in the oil recovery bottle **BEFORE** recovering the   ☐
   refrigerant.

   a. How many ounces were in the bottle? _____

6. Now hook up the A/C recovery machine to the high and low side ports. Recover the   ☐
   refrigerant from the system.

   a. How much refrigerant was recovered? _____

   b. How many ounces of oil is in the oil recovery bottle **AFTER** recovering the
   refrigerant from the system? _____

7. Describe the color of the oil recovered from the system. _____   ☐

   a. Are there any contaminants in the oil? _____

8. According to the condition of the oil, what is your opinion on the condition of the   ☐
   A/C system itself? _____

   a. What should be done? _____

9. What type of refrigerant oil is required for this system and how many ounces?   ☐

   _____

10. List where you found the service information specifying the type and oil capacity for   ☐
    this A/C system. _____

11. According to the amount of oil removed during recovery, how much new oil should   ☐
    be placed back into the system? _____

Task Completed

12. Place the proper amount of oil back into the system. ☐

13. Evacuate the A/C system for 10 minutes. Stop and observe the vacuum rating itself. ☐

   a.   What is the reading in the beginning? _____ in Hg.

   b.   Wait 10 minutes. What is the reading now? _____ in Hg.

   c.   Does the system have a leak? If so, what should be done to find the leak?

   _____

   _____

14. Continue pulling a vacuum on the system for an additional 15–20 minutes. ☐

15. After evacuating the system, recharge the system with the correct amount of refrigerant. ☐

   a.   Were there any problems with the recharge? _____

   _____

16. Ask your instructor to come and inspect your work. ☐

INSTRUCTOR'S COMMENTS _____

_____

_____

Review Questions

Name _____ Date _____ Instructor Review _____

1. The _____ includes all of the parts that produce and transfer power to the drive axle.

2. CV joint is an abbreviation for _____ _____ joint.

3. _____ mount on the frame to swivel up and down.

4. Collision technicians often complete minor mechanical repairs.
   A. True
   B. False

5. It is much easier to repair many structural parts with the powertrain removed.
   A. True
   B. False

6. The rack-and-pinion steering system is the most common type found on full-frame vehicles.
   A. True
   B. False

7. Technician A says that front-wheel drive and four-wheel drive are the same thing. Technician B says that they are different. Who is correct?
   A. Technician A
   B. Technician B
   C. Both Technician A and Technician B
   D. Neither Technician A nor Technician B

8. Technician A says that a coil spring is compressed with deadly force. Technician B says that coil springs are simply mounted and bolted to the frame. Who is correct?
   A. Technician A
   B. Technician B
   C. Both Technician A and Technician B
   D. Neither Technician A nor Technician B

9. Technician A says that air-conditioning refrigerant must be recovered. Technician B says that R-134a production and use in older vehicles has been banned. Who is correct?
   A. Technician A
   B. Technician B
   C. Both Technician A and Technician B
   D. Neither Technician A nor Technician B

10. Technician A says that the automobile industry is now using R-1234yf instead of R-134a. Technician B says that PAG oils are suitable for use in electric compressors. Who is correct?
    A. Technician A
    B. Technician B
    C. Both Technician A and Technician B
    D. Neither Technician A nor Technician B

Electrical System Operation and Service

Name _____ Date _____ Instructor Review _____

Abbreviations

Fill in the meaning of each abbreviation in the chart that follows.

Abbreviation	Meaning of Abbreviation	Abbreviation	Meaning of Abbreviation
A		POS	
AC		DTC	
OBD		LED	
BAT		SPDT	
C/B		BSD	
DC		BSM	
TAC		BSW	
MIL		V	
ALDL		W	
ABS		ADAS	
NC		AEB	
NEG		ACC	
DVOM		LDW	
LCD		LDP	

Name _____ Date _____ Instructor Review _____

Parts of an Anti-lock Brake System

Fill in the name of the part next to the appropriate number in the figure.

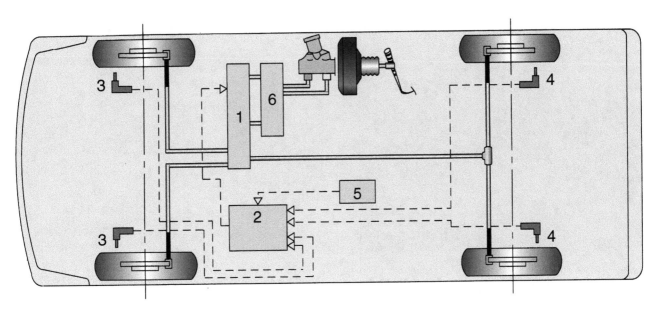

1. _____

2. _____

3. _____

4. _____

5. _____

6. _____

Name _____ Date _____ Instructor Review _____

Definitions of Terms

Define the following terms.

1. Current _____

2. Voltage _____

3. Resistance _____

4. Conductor _____

5. Series _____

6. Parallel _____

7. Ohm's Law _____

8. Magnetism _____

9. DVOM _____

10. Collision avoidance _____

11. Infinite reading _____

12. AC current _____

13. DC current _____

14. Advanced driver-assistance systems (ADAS) _____

15. Solenoid _____

16. Relay _____

17. Lidar _____

18. Ignition coil _____

19. Open circuit _____

20. Short circuit _____

21. LDP _____

22. Sensors _____

23. LDW _____

24. Scanner _____

25. Pre-repair scan _____

26. Post-repair scan _____

Name _____ Date _____

Using a Testlight and a Digital Voltmeter (DVOM)

Objective

Upon completion of this activity sheet, the student should be able to safely and correctly use a testlight and a DVOM.

ASE Education Foundation Task Correlation

III.A.1	Select and use proper personal safety equipment; take necessary precautions with hazardous operations and materials in accordance with federal, state, and local regulations. **(HP-I)**
III.A.2	Locate procedures and precautions that may apply to the vehicle being repaired. **(HP-I)**
III.A.3	Identify vehicle system hazard types (supplemental restraint system [SRS], hybrid/ electric/alternative fuel vehicles), locations, and recommended procedures before inspecting or replacing components. **(HP-I)**
III.C.1	Check for available voltage, voltage drop and current, and resistance in electrical wiring circuits and components with a DMM (digital multimeter). **(HP-I)**
III.C.2	Repair wiring and connectors. **(HP-I)**
III.C.3	Inspect, test, and replace fusible links, circuit breakers, and fuses. **(HP-I)**
III.C.4	Perform battery state-of-charge test and slow/fast battery charge. **(HP-I)**
III.C.5	Inspect, clean, repair or replace battery, battery cables, connectors, and clamps. **(HP-I)**
III.C.10	Inspect, test, and repair or replace switches, relays, bulbs, sockets, connectors, and ground wires of interior and exterior light circuits. **(HP-I)**
III.C.11	Remove and replace horn(s); check operation. **(HP-I)**
III.C.12	Check operation of wiper/washer systems; determine needed repairs. **(HP-I)**
III.C.13	Check operation of power side and tailgate window; determine needed repairs. **(HP-I)**
III.C.19	Demonstrate self-grounding procedures (anti-static) for handling electronic components. **(HP-I)**

We Support

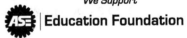

Education Foundation

Tools and Equipment

Vehicle
Testlight
DVOM
Wire strippers
Wire crimping tool
Heat shrink tubing
Soldering tool
Connector terminal tool

Safety Equipment

Safety glasses or goggles
Long-sleeved shirt

Introduction

The modern vehicle has many expensive components that keep it running. Using a testlight in preliminary troubleshooting is one of the easiest ways to check for power. The typical DVOM has three checking functions: volts, ohms, and amps.

Vehicle Description

Year _____ Make _____ Model _____

VIN _____

Battery Checks

Procedure

Task Completed ☐

1. Check to see what the voltage is, and perform a state of charge test. Google a state of charge chart, and compare the actual versus what should be.

    a. What was the actual voltage versus the recommended voltage?

    Actual _____

    Recommended _____

2. If the battery voltage was too low, it must be charged. Explain what the differences are between a slow and fast charge, and which should be done in this situation.

    _____

    _____

3. Inspect and clean the battery terminals, cables, connectors, and clamps for any damage or worn parts.

    a. List any damage found. _____

    _____

    b. Explain how you would repair or replace the battery cables, clamps, or connectors. __

    _____

    _____

Using a Testlight

Procedure

Task Completed ☐

1. Connect the alligator clip of the testlight to a good ground on the chassis.

    a. A good way to check if the bulb in the testlight itself is working is to connect the light to a good power source. Name a constant power source in the car. _____

    _____

    b. Where did you ground the light? _____

    c. Is it grounded to bare metal? _____

    d. Would this make a difference? _____

    e. If so, explain. _____

    _____

2. Now that you know the light works, place the probe at the terminal ends where the wires connect. If necessary, use paper clips to back probe connectors.   ☐

    a. Why is it not advisable to poke the probe through the coating on a wire? _____

    _____

    b. When placing the probe at the terminal, did it light up? _____

    c. If so, what does this indicate? _____

    d. If not, what does this indicate? _____

e.   If this is an indication of no power, how far up the wire in either direction do you have to go to find power? _____

3.   Reverse the process, and put the alligator clip on a known power source. Use the probe ☐
to check for ground.

a.   Where did you put the probe to complete the circuit? _____

b.   Why is it important not to place the probe in expansion terminals? _____

_____

c.   After checking different areas, check the radio. Is there constant power? _____

d.   With the key off? _____With the key on? _____

e.   Why would there be constant power to the radio if the radio is only played when the car is on? _____

f.   Check the taillights with the key on and then off. Is there constant power to the taillight? _____

g.   Is there constant power to the brake light? _____

h.   Name five different units in the vehicle that have constant power running to them.

_____

_____

4. After completing the tests on different components, answer the following questions.   ☐

a.   Do all components operate? _____

b.   If no, explain. _____

_____

c.   If any part is not operating, does it have to be replaced? _____

d.   If so, which part is it? _____

_____

Using a Digital Voltmeter (DVOM)

Procedure

5.   This exercise will use the most common DVOM with the different functions. When   ☐
measuring voltage, turn the select knob to "volts or DC."

a.   What does DC stand for? _____

b.   How is this different from AC? _____

6.   Place the black probe to the vehicle ground.   ☐

a.   Is the ground negative or positive? _____

b.   How do you know? _____

7.   Connect the red probe to the wire component to measure the voltage.   ☐

a.   What is another name for the red probe? _____

b.   What is the reading on the meter? _____

c.   Is this a normal reading? _____

d.   What does an abnormal reading mean? _____

_____

e.   What electrical part is responsible for maintaining voltage? _____

f.   Does this part need to be replaced? _____

Checking Ohms

	Task Completed
## Procedure	

8. Change the selector on your meter to ohms. If the value is not known, start at the highest range and work your way down. When using the probe be sure **NOT** to connect to any live circuit.   ☐

a.   What is the reason for starting this way? _____

b.   What is the reading on your meter? _____

c.   What does a reading of zero indicate? _____

d.   What can happen if the probe is used on a live circuit? _____

_____

e.   Does the DVOM have an internal form of protection from overload? _____

f.   If so, explain. _____

_____

g.   Using the DVOM, can you check the ohms on the speakers in the car? _____

h.   If so, how is this done? _____

_____

i.   If the wires to the speakers are run in parallel, does this change the ohms? _____

j.   If the wires are run in series, does this change the ohms? _____

k.   If the typical new vehicle from the factory has a 30-watt stereo system, what can happen if you replace it with a 300-watt subwoofer system? _____

_____

_____

Repair Damaged or Broken Wires

	Task Completed
## Procedure	

1. Use nothing but 60–40 rosin-core solder intended for electrical wiring. You will also need some PVC shrink tube.   ☐

2. Strip the wires of about ½-inch of insulation. Slip PVC shrink tube over one wire. Twist the two sections of bare wire around each other.   ☐

3. Heat the joint with a soldering iron or pencil from underneath. Apply solder to the top until molten solder wicks into the joint. Let this cool undisturbed to avoid a "cold" solder joint.   ☐

4. Heat the shrink tube to make it shrink down around the wire.   ☐

5. Use more shrink tube to bundle multiple connections.   ☐

Replacing Multiprong Connectors

Procedure

Task Completed

1. Insert the terminal tool into the connector block far enough to depress the locking tang. Wiggle the tool in a circle slightly as you pull the wire, gently to remove the connector pin from the block. Having three hands helps. ☐

2. The crimp tool has one small anvil to crimp wire directly to the metal connector pin. Once this is accomplished, use the larger anvil to crimp the strain relief over the insulated portion of the wire. ☐

3. Once it has been crimped, this leaves the connector pin ready to reinstall into the plastic block. Just push it back into the block until the tangs seat with a click. ☐

4. Practice these steps a few times before you try to do it under the dash with hot solder dripping on your cheek. That is no fun at all. ☐

Check Operation of Electrical Components

Procedure

Task Completed

1. Inspect mounting locations of and check proper operation of components such as horns, wiper/washer systems, power side windows, moveable tailgate/hatch windows, and heated mirrors. ☐

    a. How would you perform a self-grounding procedure (anti-static) for handling electronic components? _____

        _____

        _____

    b. Was the horn operating properly? If not and it is mounted properly, how would you replace it? _____

        _____

        _____

    c. Was the wiper system operating properly? If not, what is the problem and how would you correct it? _____

        _____

    d. Was the power side window and/or tailgate/hatch window operating properly? If not, what is the problem and how would you correct it? _____

        _____

        _____

    e. Was the heated mirrors operating properly? If not, what is the problem and how would you correct it? _____

        _____

        _____

INSTRUCTOR'S COMMENTS _____

_____

Name _____ Date _____

Using a Scan Tool

Objective
Upon completion of this activity sheet, the student should be able to safely and correctly use a scan tool for troubleshooting and diagnostics and also for prescan and postscan operations.

ASE Education Foundation Task Correlation

III.A.1 Select and use proper personal safety equipment; take necessary precautions with hazardous operations and materials in accordance with federal, state, and local regulations. **(HP-I)**

III.A.2 Locate procedures and precautions that may apply to the vehicle being repaired. **(HP-I)**

III.A.3 Identify vehicle system hazard types (supplemental restraint system [SRS], hybrid/electric/alternative fuel vehicles), locations, and recommended procedures before inspecting or replacing components. **(HP-I)**

III.C.7 Identify programmable electrical/electronic components, and check for malfunction indicator lamp (MIL) and fault codes; record data for reprogramming before disconnecting battery. **(HP-I)**

III.C.10 Inspect, test, and repair or replace switches, relays, bulbs, sockets, connectors, and ground wires of interior and exterior light circuits. **(HP-I)**

III.C.20 Check for module communication errors using a scan tool. **(HP-G)**

We Support

ASE | Education Foundation

Tools and Equipment
Late model vehicle
Scan tool

Safety Equipment
Safety glasses or goggles
Long-sleeved shirt

Introduction
Computers have made vehicles more efficient and safer. Computers have also made it easier to diagnose problems in the vehicle's many systems. Scan tools are expensive tools that must be used correctly to make accurate diagnoses. These can also be used for prescan and postscan operations. OEM manufacturers and insurance companies require collision centers to start making these scans a part of the everyday process for any collision repairs for most vehicles.

Vehicle Description
Year _____ Make _____ Model _____

VIN _____

Procedure
1. Scan tools can be used for prescan and postscan operations. Prescans are made before the repairs have started, so you can find and record all diagnostic trouble codes that may have been caused by the accident.

2. Postscans are made after the repairs have been completed so that you can make sure all trouble codes have been corrected and cleared off. You will plug the scan tool into the diagnostic connector for both.

3. Diagnostic connectors are in different locations depending on the vehicle. Locate the connector in your vehicle. Most are located under the dash on the driver's side. ☐

   a. What does the connector on your vehicle look like? _____

   _____

4. Using the manufacturer's requirements, plug the scanner into the connector. ☐

   a. Are there any necessary precautions to take? If so, explain. _____

   _____

5. The first test done will be the KOEO (key on, engine off) test. Turn the key to the on position. Enter the requested information for the scanner. ☐

   a. What information is requested by the scanner? _____

   _____

6. Once this information is entered, the scanner will prompt you through the next steps. ☐

   a. Are there any fault codes stored in the vehicle's memory? _____

   b. If so, what are the codes? _____

   _____

   c. If there are no fault codes in the memory, what does this indicate? _____

   _____

7. The next test will be a KOER (key on, engine running) test. ☐

   a. Is there any different information on the scanner when the engine is running? _____

   _____

   b. If so, what information is given or asked for? _____

   _____

   c. Are there any fault codes in the memory when the engine is running? _____

   d. If so, what are these codes? _____

   _____

   e. If fault codes are present, are these the same codes that showed with the engine off?

   f. If so, what does this indicate? _____

   _____

   g. If not, what does this indicate? _____

   _____

INSTRUCTOR'S COMMENTS _____

_____

_____

Review Questions

Name _____ Date _____ Instructor Review _____

1. DVOM stands for? _____

2. ADAS stands for? _____

3. A(n) _____
   carries current to parts of a circuit.

4. Ohm's Law is a math formula for calculating
   an unknown electrical value.
   A. True
   B. False

5. A series circuit has only one conductor path
   for current.
   A. True
   B. False

6. A parallel circuit has only one path for
   current.
   A. True
   B. False

7. Technician A says that a DVOM can measure
   voltage. Technician B says that a DVOM
   measures ohms. Who is correct?
   A. Technician A
   B. Technician B
   C. Both Technician A and Technician B
   D. Neither Technician A nor Technician B

8. Technician A says a postrepair scan is
   required by many manufacturers after
   collision repairs. Technician B says the
   pre-repair scan does not have to be
   performed if the 12-volt system is disabled.
   Who is correct?
   A. Technician A
   B. Technician B
   C. Both Technician A and Technician B
   D. Neither Technician A nor Technician B

9. Technician A says that windshield
   replacement may require ADAS recalibration. Technician B says that distance sensor
   removal requires ADAS recalibration. Who is
   correct?
   A. Technician A
   B. Technician B
   C. Both Technician A and Technician B
   D. Neither Technician A nor Technician B

10. Technician A says that collision avoidance
    systems use distance sensors. Technician B
    says that collision avoidance systems may
    use cameras. Who is correct?
    A. Technician A
    B. Technician B
    C. Both Technician A and Technician B
    D. Neither Technician A nor Technician B

Restraint System Operation and Service

Name _____ Date _____ Instructor Review _____

Air Bag System

Fill in the name of each part numbered on the following illustration of an air bag system.

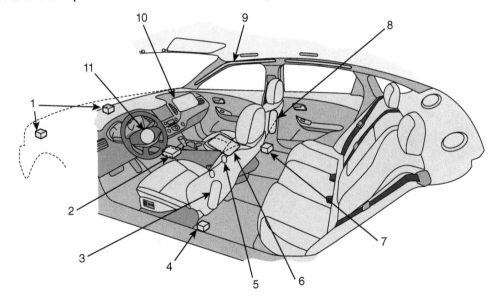

1. _____

2. _____

3. _____

4. _____

5. _____

6. _____

7. _____

8. _____

9. _____

10. _____

11. _____

Name _____ Date _____ Instructor Review _____

Passive Restraint System

Without using your textbook, fill in the names of the numbered parts in the illustration that follows.

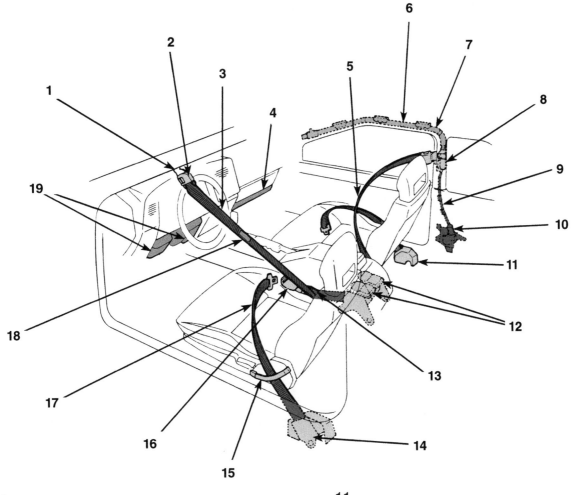

1. _____

2. _____

3. _____

4. _____

5. _____

6. _____

7. _____

8. _____

9. _____

10. _____

11. _____

12. _____

13. _____

14. _____

15. _____

16. _____

17. _____

18. _____

19. _____

Name _____ Date _____

▶Restraint System Inspection

Objective

Upon completion of this activity sheet, the student should be able to inspect, remove, and replace the components of the seat belt and shoulder harness assembly and mounting areas and verify proper operation of seat belts.

ASE Education Foundation Correlation

III.A.1	Select and use proper personal safety equipment; take necessary precautions with hazardous operations and materials in accordance with federal, state, and local regulations. **(HP-I)**
III.A.2	Locate procedures and precautions that may apply to the vehicle being repaired. **(HP-I)**
III.A.3	Identify vehicle system hazard types (supplemental restraint system [SRS], hybrid/electric/alternative fuel vehicles), locations, and recommended procedures before inspecting or replacing components. **(HP-I)**
III.A.4	Select and use a NIOSH-approved air purifying respirator. Inspect condition and ensure fit and operation. Perform proper maintenance in accordance with OSHA regulation 1910.134 and applicable state and local regulations. **(HP-I)**
III.I.1	Inspect, remove, and replace seat belt and shoulder harness assembly and components. **(HP-G)**
III.I.2	Inspect restraint system mounting areas for damage; repair as needed. **(HP-G)**
III.I.3	Inspect the operation of the seat belt system. **(HP-I)**
III.I.6	Verify that supplemental restraint system (SRS) is operational. **(HP-I)**
III.I.8	Use diagnostic trouble codes (DTC) to diagnose and repair the supplemental restraint systems (SRSs). **(HP-G)**

We Support

ASE | Education Foundation

Tools and Equipment

Vehicle
Appropriate service information
Scan tool

Safety Equipment

Safety glasses or goggles

Introduction

Vehicle restraint systems are one of the most important developments in the automotive industry. Countless lives have been saved since the advent of seat belts. It is critical that the restraint systems work as per their intended purpose in the event of a collision. This exercise will help the student learn about the proper repair procedure.

Vehicle Description

Year _____ Make _____ Model _____

VIN _____

Procedure
Seat belt Inspection

**Task
Completed**

1. Inspect the seat belt and shoulder harness webbing.   ☐

    a.   Are there any signs of damage visible in the seat belt assembly? _____

    b.   If so, describe the damage. _____

    _____

    c.   What can happen to a cut or shredded belt in an accident? _____

    _____

2. Inspect the seat belt and shoulder harness loops.   ☐

    a.   Is there any damage? If so, describe. _____

    _____

    b.   Where is the D ring located, and what is its purpose? _____

    _____

    c.   Where are the buckle attachments and anchors located? _____

    _____

3. Test the seat belt retractors for proper operation using the vehicle service information.   ☐

    a.   What is the procedure for testing the belt retractor? _____

    _____

4. Test the seat belt buckling device for proper operation.   ☐

    a.   Does the buckling device lock freely? _____

    b.   If not, what is the problem? _____

    _____

    c.   Does the buckling device release easily? _____

    d.   If not, what is the problem? _____

    _____

    e.   Why is it important that this part works correctly in an accident? _____

    _____

Air Bag Inspection

In order to find out if the air bag or SRS is working properly, we must test the system to
see if it is operational.

5. The first thing we must do is start the car. Watch the air bag light blink a few times, and   ☐
    verify that it goes off and does not stay blinking.

    a.   What does it mean if the air bag light continues to blink? _____

    _____

6. Most of the time if the air bag light goes off after the initial startup, the air bag system   ☐
    should be fully operational. There are still some cases where the air bag system may
    not be operational without it showing signs.

    a.   What would be another way to test the air bag system? _____

    _____

    _____

7. When in doubt, use a scan tool diagnostic machine and connect it to the vehicle's   ☐
    computer.

**Task
Completed**

8. Scan the computer for any air bag trouble codes, and diagnose what the problem is.   ☐

    a.   Was the air bag system operational? ⎯⎯⎯⎯⎯⎯⎯⎯⎯⎯⎯⎯⎯⎯⎯⎯⎯⎯

    b.   If not, list the trouble codes and how you would correct the problem. ⎯⎯⎯⎯⎯

    ⎯⎯⎯

    ⎯⎯⎯

INSTRUCTOR'S COMMENTS ⎯⎯⎯⎯⎯⎯⎯⎯⎯⎯⎯⎯⎯⎯⎯⎯⎯⎯⎯⎯⎯⎯⎯⎯⎯⎯⎯⎯⎯

⎯⎯

⎯⎯

Review Questions

Name _____ Date _____ Instructor Review _____

1. A(n) _____ is one that the occupants must make an effort to use.

2. A seat belt pretensioner uses a _____ to develop pressure.

3. A(n) _____ automatically deploys and inflates one or more cushions.

4. The crash sensor is the first sensor to detect a collision.
   A. True
   B. False

5. Deformation sensors are sometimes used to operate side air bags.
   A. True
   B. False

6. Multiple stage deployment means air bags can be deployed at different speeds.
   A. True
   B. False

7. Technician A says that an air bag controller analyzes inputs from a sensor to determine whether deployment is necessary. Technician B says that air bags automatically deploy at a 15 mph impact. Who is correct?
   A. Technician A
   B. Technician B
   C. Both Technician A and Technician B
   D. Neither Technician A nor Technician B

8. Technician A says that the residual powder from the air bag is an eye and skin irritant. Technician B says that the powder is not toxic. Who is correct?
   A. Technician A
   B. Technician B
   C. Both Technician A and Technician B
   D. Neither Technician A nor Technician B

9. Technician A says that seat track position sensors help determine air bag deployment. Technician B says that the air bag controller uses information from the occupant classification sensor to determine air bag inflation force. Who is correct?
   A. Technician A
   B. Technician B
   C. Both Technician A and Technician B
   D. Neither Technician A nor Technician B

10. Technician A says that a belt pretensioner is used to take up slack in a seat belt during a collision. Technician B says that pyrotechnic retractors must be replaced after air bag deployment. Who is correct?
    A. Technician A
    B. Technician B
    C. Both Technician A and Technician B
    D. Neither Technician A nor Technician B

Hybrid and Electric Vehicle Service

Name _____ Date _____ Instructor Review _____

Abbreviations

Identify hybrid terms and abbreviations, and give a brief description.

1. MSD _____
_____

2. Parallel hybrid _____
_____

3. ICE _____
_____

4. Series hybrid _____
_____

5. DVOM _____
_____

6. DTC _____
_____

7. PPE _____
_____

8. Series-parallel hybrid _____

_____

9. NiMH _____

_____

10. HVCU _____

_____

11. Regenerative braking _____

_____

12. PHEVs _____

_____

_____

Name _____ Date _____ Instructor Review _____

Hybrid Component Identification

Identify the hybrid system components from the image. Write the names of the components numbered in the image on the line next to the correct number given in the space provided.

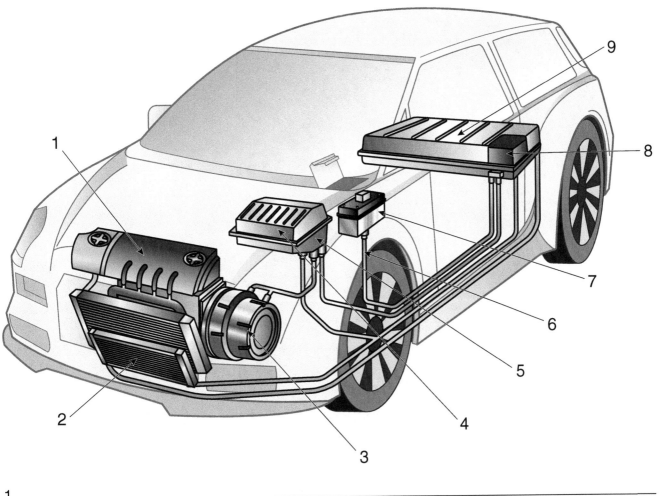

1. _____
2. _____
3. _____
4. _____
5. _____
6. _____
7. _____
8. _____
9. _____

Name _____ Date _____ Instructor Review _____

Hybrid Configuration Identification

Match the type of hybrid with the proper diagram. Write the letter of the diagram next to the correct name for that type of hybrid.

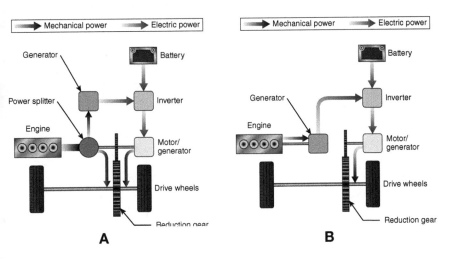
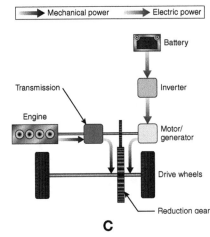

1. _____ Series hybrid
2. _____ Parallel hybrid
3. _____ Series-parallel hybrid

Name _____ Date _____

Inspecting a Hybrid Electrical System

Objective

Upon completion of this activity sheet, the student should be able to safely and accurately inspect the electrical system of a Toyota Prius Hybrid.

ASE Education Foundation Task Correlation

III.A.1 Select and use proper personal safety equipment; take necessary precautions with hazardous operations and materials in accordance with federal, state, and local regulations. **(HP-I)**

III.A.2 Locate procedures and precautions that may apply to the vehicle being repaired. **(HP-I)**

III.A.3 Identify vehicle system hazard types (supplemental restraint system [SRS], hybrid/electric/alternative fuel vehicles), locations, and recommended procedures before inspecting or replacing components. **(HP-I)**

III.A.4 Select and use a NIOSH-approved air purifying respirator. Inspect condition and ensure fit and operation. Perform proper maintenance in accordance with OSHA regulation 1910.134 and applicable state and local regulations. **(HP-I)**

III.C.6 Dispose of batteries and battery acid according to local, state, and federal requirements. **(HP-G)**

III.C.21 Use wiring diagrams, component location, and diagnostic flow charts during diagnosis of electrical circuit problems. **(HP-G)**

III.C.22 Identify safe disabling techniques of high-voltage systems on hybrid/electric vehicles. **(HP-G)**

III.C.23 Identify potential safety and materials handling concerns associated with high-voltage hybrid/electric vehicle battery systems. **(HP-G)**

III.E.13 Demonstrate an understanding of safe handling procedures associated with high-voltage A/C compressors and wiring. **(HP-G)**

III.F.2 Inspect, test, remove, and replace radiator, pressure cap, coolant system components, and water pump. **(HP-G)**

III.F.6 Demonstrate an understanding of hybrid/electric cooling systems. **(HP-G)**

III.G.7 Demonstrate an understanding of safe handling procedures associated with high-voltage powertrain components. **(HP-G)**

III.H.2 Inspect, remove, and replace fuel/DEF tank, tank filter, cap, filler hose, pump/sending unit, and inertia switch; inspect and replace fuel lines and hoses. **(HP-G)**

We Support

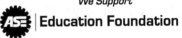 | **Education Foundation**

Tools and Equipment

Any late model hybrid
Digital volt-ohm-ammeter (DVOM)
Appropriate wrenches or sockets

Safety Equipment

Safety glasses
Insulated lineman's gloves
Rubber gloves

Introduction

Working on high-voltage electrical systems can be dangerous. Extreme caution must be taken as chances of electrocution are possible. Always work on these vehicles under the direct supervision of your instructor.

Vehicle Description

Year _____ Make _____ Model _____

VIN _____

Procedure

Task Completed

1. Locate the battery pack in the cargo area. ☐

   a. What is it mounted to and how? _____

   _____

   b. What type of battery pack is used? _____

   c. Why is the battery pack mounted in this area? _____

   _____

   d. Are there any dents or bends in the pack? _____

   e. If there is damage, what must be done with the battery pack? _____

   _____

   f. If paint is cracked, what does this indicate? _____

   _____

2. Locate the 12-volt auxiliary battery. ☐

   a. Where is it located? _____

   b. What is the purpose of this battery? _____

   c. If there is leakage, how is it repaired? _____

   _____

3. Locate the high-voltage power cables on the vehicle. ☐

   a. Where are they located? _____

   b. What color are these cables? _____

   c. What is the purpose of these color-coded cables? _____

   _____

   d. Is there any damage to the insulators? _____

   e. If there is damage to the insulators, can the cable be spliced? _____

   f. Why or why not? _____

   _____

4. Locate the inverter/converter. ☐

   a. Where is it located? _____

   b. What are the two functions of this inverter/converter? _____

   _____

   c. What is the reason this unit can be easily damaged in collision? _____

   _____

    d.   Why is the inverter/converter water cooled? Is there any damage to unit? _____

_____

    e.   If there is damage, how will it be repaired? _____

_____

    f.   If the unit has to be replaced, must it be replaced as a whole assembly? _____

_____

5. The internal combustion engine powers the vehicle and the generator.

    a.   What is the engine controlled by? _____

    b.   While checking the engine for damage, is there any damage to the fuel lines? If so, describe how it will be repaired. _____

_____

    c.   Is there any damage to the electrical wiring? If so, describe how it will be repaired.

_____

_____

    d.   Is there any damage to any pulleys? _____

    e.   If there is any damage to the pulleys, how will it be repaired? _____

_____

    f.   Is there any damage to the coolant hoses? If so, how will it be repaired? _____

_____

6. The radiator, A/C condenser, and inverter/converter are often damaged in collisions.

    a.   Why are these parts damaged so easily? _____

_____

7. Where is the A/C compressor located? _____

    a.   Can the compressor run with the engine off? If so, how? _____

_____

    b.   When checking the high-voltage cable connections, what type of tools should be used?

_____

    c.   Why is it important to use the right tools even when the HV system is disabled?

_____

8. Locate the electric motor (traction motor).   ☐

    a.   Where is it located? _____

    b.   What is the purpose of this motor? _____

    c.   When checking the transaxle housing, is there any damage present? _____

    d.   If so, how will it be repaired? _____

_____

9. Locate the engine fuel tank and fuel lines.   ☐

    a.   Where is the fuel tank located? _____

    b.   How is the tank attached? _____

_____

c. Is there any damage in the form of scratches or gauges in the fuel tank? _____

d. If so, how will the damage be repaired? _____
   _____

e. Are there any signs of bends or kinks in the fuel line? _____

f. If so, how will the fuel lines be repaired? _____
   _____

g. If the fuel lines will be replaced, why is it important to relieve the high pressure in the fuel system? _____
   _____

INSTRUCTOR'S COMMENTS _____
_____
_____

Review Questions

Name _____ Date _____ Instructor Review _____

1. ICE stands for _____
   _____.

2. HVCU stands for_____
   _____.

3. CNG stands for_____
   _____.

4. A hybrid is a vehicle that has more than one power source.
   A. True
   B. False

5. Regenerative braking is a process of recapturing kinetic energy during acceleration.
   A. True
   B. False

6. The Chevy Volt has only one power source.
   A. True
   B. False

7. Technician A says the Chevy Volt can be charged by plugging it into a 120 volt residential power supply. Technician B says the Volt can be charged by plugging it into a 240 volt residential power supply. Who is correct?
   A. Technician A
   B. Technician B
   C. Both Technician A and Technician B
   D. Neither Technician A nor Technician B

8. Technician A says class 0 gloves should be worn when pulling the high-voltage service disconnect on a hybrid. Technician B says safety glasses should be worn during the disconnect procedure. Who is correct?
   A. Technician A
   B. Technician B
   C. Both Technician A and Technician B
   D. Neither Technician A nor Technician B

9. What is the electric motor in a hybrid vehicle called?
   A. Inverter/converter
   B. Motor/generator
   C. ICE
   D. HVCU

10. Technician A says that if extreme caution is taken, it is not necessary to disable the electrical systems when welding on damaged hybrid vehicle. Technician B says that it is necessary to disable the electrical system when welding within 3 feet of the gas tank. Who is correct?
    A. Technician A
    B. Technician B
    C. Both Technician A and Technician B
    D. Neither Technician A nor Technician B

CHAPTER 25

Damage Analysis and Estimating Service

Name _____ Date _____

Single Panel Estimate

Objective

Upon completion of this activity sheet, the student should be able to write an estimate of single panel repair, using appropriate estimating software.

ASE Education Foundation Task Correlation

I.B.15   Determine the extent of the direct and indirect damage and the direction of impact; document the methods and sequence of repair. **(HP-I)**

I.B.16   Analyze and identify crush/collapse zones. **(HP-I)**

II.B.1   Review damage report, and analyze damage to determine appropriate methods for overall repair; develop and document a repair plan. **(HP-I)**

II.C.15   Identify one-time-use fasteners. **(HP-G)**

V.A.1   Select and use proper personal safety equipment; take necessary precautions with hazardous operations and materials in accordance with federal, state, and local regulations. **(HP-I)**

V.A.2   Locate procedures and precautions that may apply to the vehicle being repaired. **(HP-I)**

V.A.3   Identify vehicle system hazard types (supplemental restraint system [SRS], hybrid/electric/ alternative fuel vehicles), locations, and recommended procedures before inspecting or replacing components. **(HP-I)**

V.A.4   Select and use a NIOSH-approved air purifying respirator. Inspect condition and ensure fit and operation. Perform proper maintenance in accordance with OSHA regulation 1910.134 and applicable state and local regulations. **(HP-I)**

V.B.1   Position the vehicle for inspection. **(HP-G)**

V.B.2   Prepare vehicle for inspection by providing access to damaged areas. **(HP-G)**

V.B.3   Analyze damage to determine appropriate methods for overall repairs. **(HP-I)**

V.B.4   Determine the direction, point(s) of impact, and extent of direct, indirect, and inertia damage. **(HP-G)**

V.B.5   Gather details of the incident/accident necessary to determine the full extent of vehicle damage. **(HP-G)**

V.B.6   Identify and record preexisting damage. **(HP-I)**

V.B.7   Identify and record prior repairs. **(HP-G)**

V.B.8   Perform visual inspection of structural components. **(HP-G)**

V.B.9   Identify structural damage using measuring tools and equipment. **(HP-I)**

V.B.10   Perform visual inspection of nonstructural components. **(HP-I)**

V.B.11   Determine parts, components, material type(s), and procedures necessary for a proper repair. **(HP-I)**

V.B.12   Identify type and condition of finish; determine if refinishing is required. **(HP-I)**

V.B.13   Identify suspension, electrical, and mechanical components physical damage. **(HP-G)**

V.B.14   Identify safety systems physical damage. **(HP-G)**

V.B.15   Identify interior component damage. **(HP-I)**

V.B.16   Identify damage to add-on accessories and modifications. **(HP-G)**

V.B.17   Identify single (one-time) use components. **(HP-G)**

V.C.1   Determine and record customer/vehicle owner information. **(HP-I)**

V.C.2   Identify and record vehicle identification number (VIN) information, including nation of origin, make, model, restraint system, body type, production date, engine type, and assembly plant. **(HP-I)**

V.C.3   Identify and record vehicle mileage and options, including trim level, paint code, transmission, accessories, and modifications. **(HP-I)**

V.C.4   Identify safety systems; determine replacement items. **(HP-G)**

V.C.5   Apply appropriate estimating and parts nomenclature (terminology). **(HP-I)**

V.C.6   Determine and apply appropriate estimating sequence. **(HP-I)**

V.C.7   Utilize estimating guide procedure pages. **(HP-I)**

V.C.8   Apply estimating guide footnotes and headnotes as needed. **(HP-I)**

V.C.9   Identify operations requiring labor value judgment. **(HP-G)**

V.C.10   Select appropriate labor value for each operation (structural, nonstructural, mechanical, and refinish). **(HP-I)**

V.C.11   Select and price OEM parts; verify availability, compatibility, and condition. **(HP-G)**

V.C.12   Select and price alternative/optional OEM parts; verify availability, compatibility, and condition. **(HP-G)**

V.C.13   Select and price aftermarket parts; verify availability, compatibility, and condition. **(HP-G)**

V.C.14   Select and price recyclable/used parts; verify availability, compatibility, and condition. **(HP-G)**

V.C.15   Select and price remanufactured, rebuilt, and reconditioned parts; verify availability, compatibility, and condition. **(HP-G)**

V.C.16   Determine price and source of necessary sublet operations. **(HP-G)**

V.C.17   Determine labor value, prices, charges, allowances, or fees for non-included operations and miscellaneous items. **(HP-G)**

V.C.18	Recognize and apply overlap deductions, included operations, and additions. **(HP-I)**
V.C.19	Determine additional material and charges. **(HP-G)**
.C.20	Determine refinishing material and charges. **(HP-I)**
V.C.21	Apply math skills to establish charges and totals. **(HP-I)**
V.C.22	Identify procedural differences between computer-generated and manually written estimates. **(HP-G)**
V.C.23	Identify procedures to restore corrosion protection; establish labor values and material charges. **(HP-G)**
V.C.24	Determine the cost-effectiveness of the repair, and determine the approximate vehicle retail and repair value. **(HP-G)**
V.C.25	Recognize the differences in estimation procedures when using different information-provider systems. **(HP-G)**
V.C.26	Verify accuracy of estimate compared to the actual repair and replacement operations. **(HP-G)**
V.D.8	Identify add-on accessories. **(HP-G)**

We Support

ASE | **Education Foundation**

Tools and Equipment
Vehicle with single panel damage
Collision estimating software
Calculator
Tape measure

Safety Equipment
Safety glasses or goggles

Introduction
Knowledge of how to prepare and analyze an estimate is important for the shop manager as well as the body technician. Hidden damage that is not recognized can result in poor quality workmanship and significant financial loss for the shop and the technician.

Vehicle Description

Year _____ Make _____ Model _____

VIN _____ Paint code _____

Location of paint code _____

Procedure

Task Completed

Single Panel Repair

1. Record items such as customer information, insurance company, mileage, add-on accessories and modifications, and the VIN into the estimating software. The VIN will decode the correct information of the vehicle automatically for you if entered correctly. ☐

2. Walk around the vehicle, take note of all damage, and fill out a pre-inspection form noting prior damage that was not part of the accident. ☐

   a. Which panel are you writing an estimate for? _____

b.   Even though you are estimating only one panel, why would you note the damage on the whole vehicle? _____

_____

3. Make sure to take photos of the front, back, both sides, and from all four corners of the vehicle regardless of how small or large the damage is.   ☐

a.   Why is this important? _____

_____

4. When writing an estimate, the proper sequence is to start at the front-most panel of the accident and work your way backward to the rear of the car. If the damage occurred in the rear of the vehicle, do the opposite and work from the rear of the vehicle to the front.   ☐

Example 1: In front-end collision, start at the bumper cover, then headlights, then the hood and fenders, then doors, and so on.
Example 2: In rear collision, start at the bumper cover, then taillights, then the deck lid and quarter panels, then doors, and so on.

a.   Which area did the damage occur, and which section will you start with on the estimate? _____

_____

5. Carefully examine the dented panel. For buckled dents, use your fist to measure the size of the damage. Figure two hours for each fist that can fit in the damage.   ☐

6. Add remove and install (R&I) procedures to gain access to the backside of the panel.   ☐

a.   What panels did you include on the estimate to R&I for access? _____

_____

7. If the damage is door dings or hail damage, figure an hour for each dent.

a.   How do you estimate if there are multiple dings? _____

_____

b.   Can the dings or hail damage be repaired with paintless dent repair? _____

c.   If so, how will the estimate be different? _____

_____

8. Select the "Repair" tab, and type in the number of hours necessary to repair the panel.   ☐

9. Once the repair time is entered, it will automatically figure paint labor time.   ☐

a.   What is the number for the paint labor? _____

_____

b.   Locate OEM repair procedures for any structural repairs, panel replacement, or R&I procedures. ALLDATA has real-time updated OEM repair procedures; some can be found in the p-pages of the estimating software. A few OEM manufacturers also offer free collision repair procedures. Print these off and attach to your estimate at the end of this job sheet.

10. Apply any footnotes or headnotes to the estimate as needed.   ☐

a.   Were there any headnotes/footnotes for this estimate? If so, list them here. _____

_____

_____

11. You should always consider blending the panels on either side of the repair panel.
This ensures a good color match for the repair versus trying to paint a single panel
and hoping it matches.

   a.   What panels did you include on the estimate to blend? _____

   _____

   b.   Is there any "deduct for overlap" added to the estimate? _____

   c.   What does deduct for overlap mean? _____

   _____

12. Anytime refinish or blend has been added to an estimate, it should automatically figure
the time it takes to apply the clearcoat as well.

   a.   Why does clearcoat add time to the estimate? _____

   _____

13. Estimating software will automatically factor the body, paint, frame, and mechanical
labor for you. It will also figure in the amount of tax once you add the correct tax rate
and labor rates in the settings of the software.

   a.   What is the body rate in your area? _____

   b.   What is the paint rate in your area? _____

   c.   What is the tax rate in your area? _____

   d.   What is the estimate total now that you have everything calculated? _____

14. If you are writing a handwritten estimate, total up the body labor and paint labor.   ☐
Multiply each by the labor rate to get the dollar figures. Enter these on the appropriate
lines. Multiply the paint labor hours by the material cost per paint hour. Put this dollar
figure on the line for parts.

15. Multiply the parts costs by the tax rate. Enter this figure on the tax line. Add the lines   ☐
together for the estimate total.

16. Print off your estimate. Have your instructor go over your estimate and check your work.   ☐

INSTRUCTOR'S COMMENTS _____

_____

_____

Name _____ Date _____

Multiple Panel Estimate

Objective

Upon completion of this activity sheet, the student should be able to write an estimate of multiple panel repair, using appropriate estimating software.

ASE Education Foundation Task Correlation

I.B.15	Determine the extent of the direct and indirect damage and the direction of impact; document the methods and sequence of repair. **(HP-I)**
I.B.16	Analyze and identify crush/collapse zones. **(HP-I)**
II.B.1	Review damage report, and analyze damage to determine appropriate methods for overall repair; develop and document a repair plan. **(HP-I)**
II.C.15	Identify one-time-use fasteners. **(HP-G)**
V.A.1	Select and use proper personal safety equipment; take necessary precautions with hazardous operations and materials in accordance with federal, state, and local regulations. **(HP-I)**
V.A.2	Locate procedures and precautions that may apply to the vehicle being repaired. **(HP-I)**
V.A.3	Identify vehicle system hazard types (supplemental restraint system [SRS], hybrid/electric/ alternative fuel vehicles), locations, and recommended procedures before inspecting or replacing components. **(HP-I)**
V.A.4	Select and use a NIOSH-approved air purifying respirator. Inspect condition and ensure fit and operation. Perform proper maintenance in accordance with OSHA regulation 1910.134 and applicable state and local regulations. **(HP-I)**
V.B.1	Position the vehicle for inspection. **(HP-G)**
V.B.2	Prepare vehicle for inspection by providing access to damaged areas. **(HP-G)**
V.B.3	Analyze damage to determine appropriate methods for overall repairs. **(HP-I)**
V.B.4	Determine the direction, point(s) of impact, and extent of direct, indirect, and inertia damage. **(HP-G)**
V.B.5	Gather details of the incident/accident necessary to determine the full extent of vehicle damage. **(HP-G)**
V.B.6	Identify and record preexisting damage. **(HP-I)**
V.B.7	Identify and record prior repairs. **(HP-G)**
V.B.8	Perform visual inspection of structural components. **(HP-G)**
V.B.9	Identify structural damage using measuring tools and equipment. **(HP-I)**
V.B.10	Perform visual inspection of nonstructural components. **(HP-I)**
V.B.11	Determine parts, components, material type(s), and procedures necessary for a proper repair. **(HP-I)**
V.B.12	Identify type and condition of finish; determine if refinishing is required. **(HP-I)**
V.B.13	Identify suspension, electrical, and mechanical components physical damage. **(HP-G)**
V.B.14	Identify safety systems physical damage. **(HP-G)**
V.B.15	Identify interior component damage. **(HP-I)**

V.B.16   Identify damage to add-on accessories and modifications. **(HP-G)**

V.B.17   Identify single (one-time) use components. **(HP-G)**

V.C.1   Determine and record customer/vehicle owner information. **(HP-I)**

V.C.2   Identify and record vehicle identification number (VIN) information, including nation of origin, make, model, restraint system, body type, production date, engine type, and assembly plant. **(HP-I)**

V.C.3   Identify and record vehicle mileage and options, including trim level, paint code, transmission, accessories, and modifications. **(HP-I)**

V.C.4   Identify safety systems; determine replacement items. **(HP-G)**

V.C.5   Apply appropriate estimating and parts nomenclature (terminology). **(HP-I)**

V.C.6   Determine and apply appropriate estimating sequence. **(HP-I)**

V.C.7   Utilize estimating guide procedure pages. **(HP-I)**

V.C.8   Apply estimating guide footnotes and headnotes as needed. **(HP-I)**

V.C.9   Identify operations requiring labor value judgment. **(HP-G)**

V.C.10   Select appropriate labor value for each operation (structural, nonstructural, mechanical, and refinish). **(HP-I)**

V.C.11   Select and price OEM parts; verify availability, compatibility, and condition. **(HP-G)**

V.C.12   Select and price alternative/optional OEM parts; verify availability, compatibility, and condition. **(HP-G)**

V.C.13   Select and price aftermarket parts; verify availability, compatibility, and condition. **(HP-G)**

V.C.14   Select and price recyclable/used parts; verify availability, compatibility, and condition. **(HP-G)**

V.C.15   Select and price remanufactured, rebuilt, and reconditioned parts; verify availability, compatibility, and condition. **(HP-G)**

V.C.16   Determine price and source of necessary sublet operations. **(HP-G)**

V.C.17   Determine labor value, prices, charges, allowances, or fees for non-included operations and miscellaneous items. **(HP-G)**

V.C.18   Recognize and apply overlap deductions, included operations, and additions. **(HP-I)**

V.C.19   Determine additional material and charges. **(HP-G)**

V.C.20   Determine refinishing material and charges. **(HP-I)**

V.C.21   Apply math skills to establish charges and totals. **(HP-I)**

V.C.22   Identify procedural differences between computer-generated and manually written estimates. **(HP-G)**

V.C.23   Identify procedures to restore corrosion protection; establish labor values and material charges. **(HP-G)**

V.C.24   Determine the cost-effectiveness of the repair, and determine the approximate vehicle retail and repair value. **(HP-G)**

V.C.25   Recognize the differences in estimation procedures when using different information-provider systems. **(HP-G)**

V.C.26   Verify accuracy of estimate compared to the actual repair and replacement operations. **(HP-G)**

V.D.8   Identify add-on accessories. **(HP-G)**

Tools and Equipment

Vehicle with multiple damaged panels
Collision estimating software
Calculator
Measuring tools and equipment

Safety Equipment

Safety glasses or goggles

Introduction

Knowledge of how to prepare and analyze an estimate is important for the shop manager as well as the body technician. Hidden damage that is not recognized can result in poor quality workmanship and significant financial loss for the shop and the technician.

Vehicle Description

Year _____ Make _____ Model _____

VIN _____ Paint code _____

Location of paint code _____

Procedure

Multiple Panel Repair

Task Completed

1. Call one independent body shop and one dealership body shop to find out their hourly labor rate. Use the same amount for paint labor. Remember that labor rates can vary from shop to shop as some choose to set lower rates than others to try to get more work in the shop.   ☐

   a.   The independent shop rate is _____ per hour.

   b.   The dealership rate is _____ per hour.

2. Check the whole vehicle for damage, not just the apparent damage. Complete a pre-inspection form, and review it with the customer. Have them sign it as this will help eliminate any problems if the customer says damage to the vehicle happened while in the shop.   ☐

   a.   Before writing any of the estimate, describe the damage on the vehicle and explain the angle of impact of the accident. _____ _____ _____

   b.   Can any damaged panels be repaired, or must they be replaced? _____ _____

3. Now go over each panel entirely being mindful not to forget any categories listed for each part group or section. Go over these steps with each panel that is damaged or needs to be blended. Be thorough and look for forgettable items.   ☐

4. If the panel is to be replaced, select the "Replace" tab.

   a.   If replaced, how many hours are given to replace the panel? _____

   b.   If replaced, how many hours are given to paint the panel? _____

   c.   Is there any overlap time? _____

   d.   What is the reason for the overlap deduction? _____ _____

   e.   If the panel is to be repaired, how is this determined? _____ _____

5. If repaired, select the "Repair" tab and enter the number of hours necessary to repair it.   ☐

    a.   Are the hours to repair the panel more than to replace it? _____

    b.   If so, explain why you would repair it. The general rule of thumb is if the amount to repair it is more than 75 percent of the price of a new panel, replace it.

    _____

    _____

6. Check for structural damage both visually and with measuring tools and equipment. Record your findings, and add structural or frame repair to the estimate for all the areas you found damage.   ☐

    a.   What tools did you use to measure? _____

    b.   Was there any structural or frame damage? If so, list here what is bent. _____

    _____

    _____

7. Look in the passenger compartment for any interior damage.   ☐

    a.   Was there any interior damage? If so, list the damage panels. _____

    _____

8. Locate OEM repair procedures for any structural repairs, panel replacement, or R&I procedures. ALLDATA has real-time updated OEM repair procedures; some can be found in the p-pages of the estimating software. A few OEM manufacturers offer free collision repair procedures also. Print these off and attach to your estimate at the end of this job sheet.   ☐

9. Apply any footnotes or headnotes to the estimate as needed.   ☐

    a.   Were there any headnotes/footnotes for this estimate? If so, list them here. ____

    _____

    _____

10. Many panels have pinstripes, moldings, emblems, and nameplates that get overlooked in an estimate. Examine the vehicle to see whether any parts like these must be replaced.   ☐

    a.   On your panels, do any other parts need to be replaced? _____

    b.   If so, what parts are they? _____

    _____

    c.   How much time is allowed to replace this part? _____

11. Once you are confident that you have not missed any damage, check again. Make one last complete and thorough walk through before you are done. There are normally one or two hidden or forgotten items.   ☐

    a.   What is the grand total? _____

12. Print off your estimate. Have your instructor go over your estimate and check your work.   ☐

INSTRUCTOR'S COMMENTS _____

_____

Name _____ Date _____

Customer Relations Skills

Objective

Upon completion of this activity sheet, the student should be able to write an estimate of vehicle repair, using a crash estimating guide.

ASE Education Foundation Task Correlation

V.A.2   Locate procedures and precautions that may apply to the vehicle being repaired. **(HP-I)**

V.B.5   Gather details of the incident/accident necessary to determine the full extent of vehicle damage. **(HP-G)**

V.B.6   Identify and record preexisting damage. **(HP-I)**

V.B.7   Identify and record prior repairs. **(HP-G)**

V.B.10 Perform visual inspection of nonstructural components. **(HP-I)**

V.C.1   Determine and record customer/vehicle owner information. **(HP-I)**

V.E.1   Acknowledge and/or greet customer/client. **(HP-I)**

V.E.2   Listen to customer/client; collect information and identify customers/client's concerns, needs, and expectations. **(HP-I)**

V.E.3   Establish cooperative attitude with customer/client. **(HP-I)**

V.E.4   Identify yourself to customer/client; offer assistance. **(HP-I)**

V.E.5   Deal with angry customer/client. **(HP-I)**

V.E.6   Identify customer/client preferred communication method; follow up to keep customer/client informed about parts and the repair process. **(HP-G)**

V.E.8   Project positive attitude and professional appearance. **(HP-I)**

V.E.9   Provide and review warranty information. **(HP-I)**

V.E.11 Estimate and explain duration of out-of-service time. **(HP-G)**

V.E.13 Interpret and explain manual or computer-assisted estimate to customer/client. **(HP-I)**

We Support

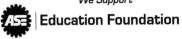 **Education Foundation**

Tools and Equipment

Customer needing repairs
Estimate of repair
Positive and cooperative attitude

Safety Equipment

Safety glasses or goggles

Introduction

Knowledge of how to properly communicate with the customer, and to make them feel as comfortable as possible, is important for the shop manager as well as the body technician. The customer has probably just been in an accident that has them rattled, scared, and/or upset. You want them to feel comfortable, like they will be treated like family and not taken advantage of. Be honest and up-front with them every step of the way. If the customer feels that you will treat them with respect and look out for their best interests, that means as much or more to them as performing a thorough inspection and repair.

Vehicle Description

Year _____ Make _____ Model _____

VIN _____

Procedure

Greeting the Customer

1. When you are an estimator or a customer relations specialist, you must always make
   sure that you are dressed professionally and proper. Make sure that your hair is fixed,
   polo or button up shirt is tucked in and neat, and there are no clothes with holes or rips
   in them. ☐

   a. Why is it important to dress and look nice for a customer? _____
      _____
      _____

2. Introduce yourself to the customer when they walk in or up to you. Make eye contact
   with the customer, smile, acknowledge, and shake the customer's hand when they walk
   up to you. The first impression is often what sticks in the customer's mind, so be sure
   they remember a smiling face and a courteous person when they think of you. ☐

3. First, you must establish a cooperative attitude with the customer. Most of the time a
   customer comes to you because they have been in an accident and they are either
   nervous, scared still from the accident, or upset and angry because most of the time, in
   their eyes the wreck was not their fault. ☐

   a. What was the cause of the customer's accident? _____
      _____
      _____

4. If they are still angry about what happened or get upset about anything during the
   conversation, the best way to calm them down is to steer them away from the topic that
   is making them upset. Ask them if they would like a bottle of water or a cup of coffee.
   Reassure them that you will do whatever it takes to make sure that they are completely
   satisfied and happy with the work that you will provide them. Remember, they are scared
   during this process, sometimes a little reassurance for an upset customer is all it takes to
   calm them down. ☐

   a. What seemed to upset them the most, and how did you calm them down? _____
      _____
      _____

5. Gather as much information as you can about how the accident happened and what they
   would like you to do for them. Listening to the customer is one of the most important
   things you can do for them because it makes them feel that you really care. Make sure to
   ask if they have any questions about the repair process or concerns or needs that you
   can answer for them. ☐

   a. Did the customer have any concerns, needs, or expectations about the repair?
      _____

   b. If so, list them here. _____
      _____

6. Make sure to go over the estimate with the customer. Explain to them what you are
   going to do to their car so that they are comfortable and not concerned or worried about
   what will happen next. By doing this, it allows them to have a sense of control and
   satisfaction. ☐

7. You should have noticed their attitude change and that they have calmed down since they first came to you.

a. If so, how are they acting now? _____

_____

_____

8. Provide and explain to them any warranty information that your shop may offer. If you offer a lifetime paint warranty, parts warranty, or repair warranty, make sure to explain to them how it all works. Also, any technical or consumer protection information that may be applicable. ☐

a. What kind of warranties do you offer for the customer? _____

_____

b. How did they seem to feel after explaining this to them? _____

_____

9. Once you have obtained a mutual agreement with the customer to repair their vehicle, ask them their preferred communication method. This means whether they would rather you call them at their work or cell phone, if they would rather be e-mailed or notified through text message, or all the above. You also want to ask them what time of day is best for them to be reached at by their preferred method of communication. ☐

a. What was their preferred method of communication and best time available? _____

_____

10. Walk around the vehicle with the customer, and take note of all prior damage and all accident-related damage. Print and fill out a pre-inspection form noting prior damage that was not part of the accident and have the customer review and sign the inspection form. Attach the pre-inspection form you filled out with them to this job sheet. ☐

11. Make sure to follow up with the customer's needs and expectations throughout the repair process. This means to let them know when you have received all the correct parts for their vehicle, when you have completed the body repair process, and again when entering the paint process and for reassembling the vehicle. Let them know at least a day or two ahead of time for when you believe the car will be finished and ready to pick up. Sometimes things happen in the repair process and you may not meet this deadline, or you may be done sooner. ☐

12. Proper communication is the absolute most important thing with the customer during the repair process. Always keep them updated so they are not left in the dark with questions because you have not kept in touch with them or explained enough to them.

INSTRUCTOR'S COMMENTS _____

_____

_____

Review Questions

Name _____ Date _____ Instructor Review _____

1. A _____ is a printed form that outlines the work to be completed by the technician.

2. The _____ is a preset amount of time and money charged for a specific repair.

3. Blueprinting requires a disassembly of all the affected areas of the vehicle.
   A. True
   B. False

4. Crash estimating guides can be used as a reference for pricing parts.
   A. True
   B. False

5. 0.3 of an hour is equal to 30 minutes.
   A. True
   B. False

6. The basic paint time generally includes time for tinting the paint to match.
   A. True
   B. False

7. Technician A says that you can use an iPad, tablet, and your phone to take pictures of the vehicle. Technician B says these can also be used to load the vehicle information. Who is correct?
   A. Technician A
   B. Technician B
   C. Both Technician A and Technician B
   D. Neither Technician A nor Technician B

8. Technician A says that an insurance adjuster works only for the insurance company. Technician B says that an insurance adjuster works for the collision shop. Who is correct?
   A. Technician A
   B. Technician B
   C. Both Technician A and Technician B
   D. Neither Technician A nor Technician B

9. Technician A says that a door trim panel is the trim molding on the outside of the door. Technician B says that the trim panel is on the inside of the door. Who is correct?
   A. Technician A
   B. Technician B
   C. Both Technician A and Technician B
   D. Neither Technician A nor Technician B

10. Technician A says that all pinstripes are taped on. Technician B says that all pinstripes are painted on. Who is correct?
    A. Technician A
    B. Technician B
    C. Both Technician A and Technician B
    D. Neither Technician A nor Technician B

CHAPTER 26

Refinishing Equipment

Name _____ Date _____ Instructor Review _____

Paint Gun Parts Identification

Fill in the name of each part numbered on the following illustration of a paint gun.

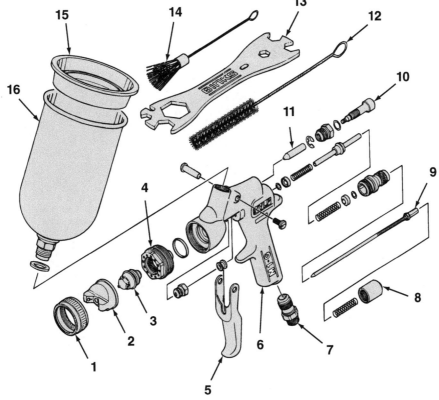

1. _____

2. _____

3. _____

4. _____

5. _____

6. _____

7. _____

8. _____

9. _____

10. _____

11. _____

12. _____

13. _____

14. _____

15. _____

16. _____

Name _____ Date _____ Instructor Review _____

Paint Gun Use

Without using your textbook, fill in the name and use for each paint gun shown in the illustration that follows. Explain how each system works.

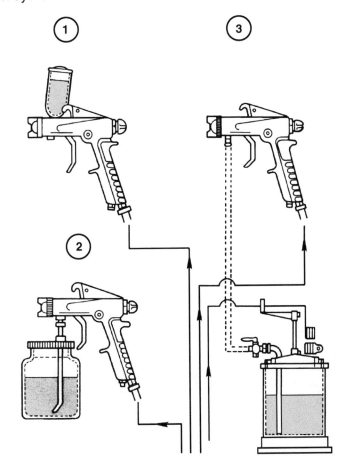

1. _____ Use _____

   How the system works: _____

   _____

2. _____ Use _____

   How the system works: _____

   _____

3. _____ Use _____

   How the system works: _____

   _____

Name _____ Date _____ Instructor/Review _____

Paint Gun Use

Name _____ Date _____ Instructor Review _____

Paint Gun Identification

Label the image with the correct term for each part of the paint gun

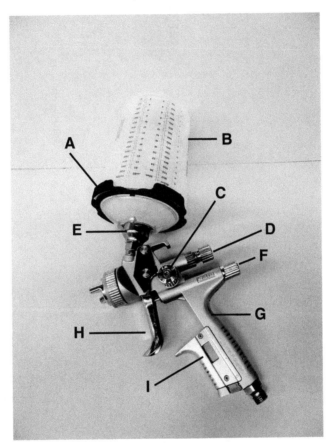

A. _____   F. _____

B. _____   G. _____

C. _____   H. _____

D. _____   I. _____

E. _____

Name _____   Date _____

Refinishing Equipment Setup

Objective

Upon completion of this activity sheet, the student should be able to inspect and safely operate the paint equipment in any collision center.

ASE Education Foundation Task Correlation

IV.A.3   Inspect spray environment and equipment to ensure compliance with federal, state, and local regulations and for safety and cleanliness hazards. **(HP-I)**

IV.A.4   Select and use a NIOSH-approved air purifying respirator. Inspect condition and ensure fit and operation. Perform proper maintenance in accordance with OSHA regulation 1910.134 and applicable state and local regulations. **(HP-I)**

IV.A.5   Select and use a NIOSH-approved supplied air (fresh air make-up) respirator system. Perform proper maintenance in accordance with OSHA regulation 1910.134 and applicable state and local regulations. **(HP-I)**

IV.A.6   Select and use the proper personal safety equipment for surface preparation, spray gun and related equipment operation, paint mixing, matching and application, paint defects, and detailing (gloves, suits, hoods, eye and ear protection, etc.). **(HP-I)**

We Support

ASE | Education Foundation

Tools and Equipment

Paint guns
Collision center paint equipment
(Most will have to be located
individually for this assignment.)

Safety Equipment

Safety glasses or goggles

Introduction

Paint jobs are a costly part of the collision repair procedures. Today's paint jobs are often guaranteed for the life of the vehicle. To produce quality paint jobs, good material, skilled labor, and clean, efficient paint equipment are necessary. This job sheet will help you become familiar with the different equipment necessary to produce a quality paint job.

   If you have any difficulty or doubt about any aspect of this activity, be sure to ask your instructor for help.

Procedure

1.   There are three main types of spray guns: suction-, gravity-, and pressure-fed.

   a.   What type of paint guns does your shop have? _____

   b.   What brand are the guns? _____

   c.   Which guns are used for primer-surfacer? _____

   d.   What determines whether a gun is used for primer? _____

   e.   Which guns are used for paint and clearcoat? _____

Task Completed

2. Most modern collision centers have their own paint-mixing machines. Mixing machines save the shop money and produce less waste because the painter can mix just enough paint for a particular job. For safety reasons, mixing should be in an enclosed room separate from, but connected to, the paint booth. Locate your shop's mixing machine. ☐

    a. What brand of paint is on the machine? _____

    b. According to your instructor, why was this brand of paint chosen? _____

    _____

    c. Why is it important that the mixing room be separate from the paint booth? _____

    _____

    _____

    d. Is a digital scale used to weigh mixing tints?

    _____

3. Locate your paint disposal (hazardous waste) system. ☐

    a. Where is it located? _____

    b. Is it vented? If so, where? If not, why not? _____

    _____

    _____

    c. How is this container disposed of when full? _____

    _____

    _____

4. Locate the paint booth in your shop. ☐

    a. Where is it located? _____

    b. What kind of booth is it? _____

    c. How does a downdraft booth operate? _____

    _____

    d. How does a semi-downdraft booth operate? _____

    _____

    e. How does a crossflow booth operate? _____

    _____

    f. Is your booth equipped with an oven? _____

    g. How long does it take to bake the paint on a vehicle? _____

    h. Is there a fire sprinkling system in the booth? _____

    i. Does the booth have a wet or dry filtration system? _____

5. Respirators are required when painting to keep airborne solvents and particles from being inhaled. Even with the use of a paint booth, quality respirators are required. Get your own respirators out to answer the following. ☐

    a. What type of respirators do you have and use? _____

    b. What will each of these types of respirators protect you from? _____

    _____

    c. How can you check the fit of the respirator doing a negative pressure test? _____

    _____

**Task
Completed**

d. How do you perform a positive pressure test? _____

e. What is a good way to tell when cartridges need replacement? _____

_____

6. The safest way to spray is using an air-supplied respirator. Though more costly, a supplied air system gives the painter the maximum protection.

a. Is your shop equipped with a supplied air system? _____

b. Does this system use a half-mask or full-mask respirator? _____

c. How does a fresh air system work? _____

_____

7. Locate the shop gun cleaner. These gun cleaners use thinner that is continuously being recycled while the gun is in operation. It is important that a paint gun be emptied properly **BEFORE** being cleaned in the gun cleaner. ☐

a. How does this system work? _____

_____

b. What is the reason for having two different thinners in the cleaner? _____

_____

c. When the cleaning cycle is complete, where are the paint guns stored? _____

_____

INSTRUCTOR'S COMMENTS _____

_____

_____

Name _____ Date _____

Paint Gun Cleaning and Setup

Objective

Upon completion of this activity sheet, the student should be able to inspect, disassemble, clean, and reassemble any spray gun within the time allotted by the instructor.

ASE Education Foundation Task Correlation

IV.A.4 Select and use a NIOSH-approved air purifying respirator. Inspect condition and ensure fit and operation. Perform proper maintenance in accordance with OSHA regulation 1910.134 and applicable state and local regulations. **(HP-I)**

IV.A.5 Select and use a NIOSH-approved supplied air (fresh air make-up) respirator system. Perform proper maintenance in accordance with OSHA regulation 1910.134 and applicable state and local regulations. **(HP-I)**

IV.A.6 Select and use the proper personal safety equipment for surface preparation, spray gun and related equipment operation, paint mixing, matching and application, paint defects, and detailing (gloves, suits, hoods, eye and ear protection, etc.). **(HP-I)**

IV.C.1 Inspect, clean, and determine condition of spray guns and related equipment (air hoses, regulators, air lines, air source, and spray environment). **(HP-I)**

IV.C.2 Select spray gun setup (fluid needle, nozzle, and cap) for product being applied. **(HP-I)**

IV.C.3 Test and adjust spray gun using fluid, air, and pattern control valves. **(HP-I)**

IV.C.4 Demonstrate an understanding of the operation of pressure spray equipment. **(HP-G)**

We Support

ASE | Education Foundation

Tools and Equipment

Paint gun
Gun cleaning solvents
Masking tape
Tape measure

Safety Equipment

Safety glasses or goggles
Paint respirator
Long-sleeved shirt
Rubber gloves

Introduction

Spray guns are expensive, precision equipment that must be cleaned and maintained properly to ensure years of quality performance. Although there are many different types and brands of spray guns, their cleaning and maintenance are the same. This job sheet will help you learn the necessary steps to clean and maintain any given spray gun.

   If you have any difficulty or doubt about any aspect of this activity, be sure to ask your instructor for help.

Gun Description

Note the following data on the specific spray gun you are working with during this job.

Brand _____

Type _____

Procedure

1. It is important to know of the three types of spray guns: suction-fed, gravity-fed, and pressure-fed. They each serve different purposes. How do these different types of guns work, and how and when are they used?

   a. Suction-fed _____

   _____

   b. Gravity-fed _____

   _____

   c. Pressure-fed _____

   _____

2. Mix primer or basecoat to check the function of the spray gun. On large areas of masking paper, spray the paint in the recommended procedure.   ☐

   a. Is the spray pattern adjustment working properly? If not, explain. _____

   _____

   b. If the gun is working properly, open the pattern as wide as possible and spray the paper. What is the height of the pattern? _____

   c. Close the pattern to the smallest possible pattern and spray. What is the height of that pattern? _____

   d. Is the fluid adjustment knob working properly? _____

   e. How did you determine this? _____

   _____

   f. What is the trigger action with the knob all the way in? _____

   _____

   g. What is the trigger action with the knob all the way out? _____

   _____

3. Once the gun is spraying properly, cover the air vent on the top of the cap and spray.   ☐

   a. How is the gun spraying now? _____

   _____

4. Clear the vent hole to return spray function. Cover the hole on the right side of the horn with masking tape.   ☐

   a. How is the spray pattern affected? _____

   _____

   b. Repeat with the left side. How is the pattern affected with both holes covered? _____

   _____

5. Remove the cup or PPS cup from the gun body.   ☐

   a. Is the cup on the top or bottom of the gun? _____

   b. This would make the gun a _____ -fed gun.

6. Remove the air cap from the gun. This cap is only hand-tightened and should unscrew easily.   ☐

   a. Are there any numbers or letters on the cap? If so, what are they? _____

   _____

   b. Explain what these numbers or letters mean. _____

   _____

c.   How many holes are in each horn of the cap? _____

d.   What is the function of these holes? _____

_____

e.   How many holes are in the auxiliary orifice? _____

f.   What is the function of these holes? _____

_____

g.   What is the function of the center hole? _____

_____

7. Soak the part in a mixing cup filled with gun cleaning solvent. This will help break off any ☐
dried material. All the holes in the spray gun parts are precisely machined and must not
be altered using any tool that can distort their size. Let the material do the job it was
designed for and all the parts will clean easily.

8. With the proper wrench, remove the needle and then the fluid tip. ☐

a.   Is there a number on the fluid tip? _____

b.   If so, what is it and what does it represent? _____

_____

9. Examine the threads on the fluid tip to ensure that none are stripped and will reinstall ☐
easily.

10. Remove the fluid needle. ☐

a.   What is the purpose of the fluid needle? _____

_____

Cleaning and Reassembly

11. It is best to use a spray gun cleaning solvent on any paint gun. The solvent will easily ☐
remove paint deposits without damaging delicate internal gun parts. Every shop should
be equipped with a lubricant especially designed for spray guns. If any other oils are
used, the paint can become contaminated and cause paint problems. Using an acid
brush or any other small brush, clean each small part, concentrating on all holes. As
you clean each part, lay the part out on a clean workbench.

12. Before reinstalling the chrome head to the gun body, make sure all the holes are cleaned ☐
carefully. A gun is usually not disassembled to this degree, so take advantage and
thoroughly clean every part and orifice. An Allen screw cannot be started by hand, so
it is important that the proper tool be used.

13. Install the fluid tip, and secure it with the proper wrench. ☐

a.   What is the size of the head of the fluid tip? _____

b.   Is this part lubricated? _____

14. Reinstall the head onto the gun body, and tighten with the gun wrench. ☐

15. Reinstall the air cap. Make sure the horns are horizontal. ☐

a.   How will the pattern be affected if the horns are vertical? _____

_____

16. Reinstall the cup. ☐

Task Completed

17. Mix paint or primer and test the operation of the spray gun. Get your instructor to watch   ☐
the test patterns now with you so they can check your work.

a.   Does the gun spray properly? If not, explain. _____

_____

b.   How will you correct this problem? _____

_____

18. Continue to test until the gun is operating correctly.   ☐

INSTRUCTOR'S COMMENTS _____

_____

_____

Review Questions

Name _____ Date _____ Instructor Review _____

1. The second stage of atomization occurs when the paint stream is struck by jets of air from the air cap horns.
   A. True
   B. False

2. The _____ adjusts the fan size.

3. Technician A uses a gravity-fed gun for primer. Technician B uses a siphon for primer. Who is correct?
   A. Technician A
   B. Technician B
   C. Both Technician A and Technician B
   D. Neither Technician A nor Technician B

4. If the air valve sticks, the trigger will no longer control the air flow.
   A. True
   B. False

5. If a siphon-fed gun does not spray properly, the first item you should check is the cup vent hole.
   A. True
   B. False

6. When mixing in a PPS cup, the mixing instructions for the most relevant ratios are located on the liners.
   A. True
   B. False

7. Viscosity is measured with a Chevy cup.
   A. True
   B. False

8. The proper way to adjust a paint gun is to set the fan, followed by _____, then _____.

9. _____ means the painter is not holding the gun perpendicular to the surface.

10. Technician A says that solvent pop is a common problem with HVLP guns. Technician B believes that HVLP paint guns save on material. Who is correct?
   A. Technician A
   B. Technician B
   C. Both Technician A and Technician B
   D. Neither Technician A nor Technician B

Surface Preparation and Masking

Name _____ Date _____ Instructor Review _____

Using Different Grits of Sandpaper

Without using your textbook, fill in the use for each of the different grits of sandpaper.

Grit	Aluminum Oxide	Silicon Carbide	Zirconia Alumina	Primary Use for Auto Body Repair
1.	—	5,000	—	
2.	—	3,000	—	
3.	—	1,500	—	
4.	—	800	—	
5.	—	600	600	
6.	400	400	400	
	320	320	—	
	280	280	280	
	240	240	240	
7.	220	220	—	
8.	180	180	180	
	150	150	150	
9.	120	120	—	
	100	100	100	
	80	80	80	
10.	60	60	60	
	50	50	—	
	40	40	40	
	36	36	—	
11.	24	24	24	
	16	16	—	

1. _____

2. _____

3. _____

4. _____

5. _____

6. _____

7. _____

8. _____

9. _____

10. _____

11. _____

Name _____ Date _____ Instructor Review _____

Definitions of Materials and Techniques

Without using your textbook, define the following materials and techniques.

1. Surface preparation _____

2. Film breakdown _____

3. Plastic media _____

4. Metal conditioner _____

5. Shipping coating _____

6. E-coat _____

7. Self-etching primer _____

8. Seam sealer _____

9. Sealer _____

10. Primer _____

11. Primer-sealer _____

12. Primer filler _____

13. Adhesion promoter _____

14. Flash time _____

15. Compounding _____

16. Spot putty _____

17. Bull's-eye _____

18. Masking _____

19. Double masking _____

20. Back masking _____

Name _____ Date _____

Prepping and Featheredging

Objective

Upon completion of this activity sheet, the student should be able to remove trim, clean, and prep a vehicle for refinishing and featheredge repair areas.

ASE Education Foundation Task Correlation

II.B.1	Review damage report, and analyze damage to determine appropriate methods for overall repair; develop and document a repair plan. **(HP-I)**
II.B.2	Inspect, remove, label, store, and reinstall exterior trim and moldings. **(HP-I)**
IV.B.1	Inspect, remove, store, protect, and replace exterior trim and components necessary for proper surface preparation. **(HP-I)**
IV.B.2	Soap and water wash entire vehicle; use appropriate cleaner to remove contaminants. **(HP-I)**
IV.B.3	Inspect and identify type of finish, surface condition, and film thickness; develop and document a plan for refinishing using a total product system. **(HP-G)**
IV.B.4	Remove paint finish as needed. **(HP-I)**
IV.B.5	Dry or wet sand areas to be refinished. **(HP-I)**
IV.B.6	Featheredge areas to be refinished. **(HP-I)**
IV.B.16	Remove dust from area to be refinished, including cracks or moldings of adjacent areas. **(HP-I)**
IV.B.17	Clean area to be refinished using a final cleaning solution. **(HP-I)**

We Support

ASE | **Education Foundation**

Tools and Equipment

Vehicle or panel to be refinished
Trim removal tools if necessary
Appropriate solvents: soap, wax and grease remover, and lacquer thinner
DA sander
Hand block sander (or flex block)
Appropriate grit sanding discs

Safety Equipment

Safety glasses or goggles
Protective gloves
Appropriate dust mask

Introduction

It has been said repeatedly that paint preparation is 95 percent of a quality paint job. Even the best brand of paint will not adhere to a panel if the panel is not prepared properly. With the increasing cost of automobiles, it is important that paint jobs be restored to OEM standards for paint performance purposes and to retain its retail value.

Vehicle Description

Year _____ Make _____ Model _____

VIN _____

Procedure

Task Completed

1. What is the general extent of the damage to the vehicle? _____
_____
_____

   a. Which panels are damaged? _____

2. Examine the moldings. ☐

   a. If any of the moldings need to be removed, what tool is needed to remove these moldings? _____

3. Once the moldings are removed, mark them and store them where they will not be damaged. ☐

4. Wash the entire vehicle with the appropriate soap and water. ☐

   a. If the vehicle is to be sanded anyway, what is the purpose of first washing it? _____
   _____

5. Using a paint film thickness gauge, measure the average film thickness of the area to be repaired. ☐

   a. What is the thickness of this panel/vehicle you are working on? _____

   b. What is the average mil thickness from OEM? _____

6. Assuming the damaged area has been repaired to your instructor's approval, begin the featheredging procedure. ☐

   a. What is featheredging, and how is it done? _____
   _____

   b. What grit sandpapers are used for this job? _____

   c. How far should the finish be featheredged? _____

   d. If not properly featheredged, what will appear in the paint? _____
   _____

7. With your instructor's approval, blow off the panel and clean with a prepping solvent making sure you do not allow it to air dry. You must continue wiping until completely evaporated. ☐

   a. What can happen if this solvent is not used? _____
   _____

   b. Why is it important that the solvent not be allowed to dry on the panel? _____
   _____

8. Tack the panel off free from any lint or dust particles. ☐

INSTRUCTOR'S COMMENTS _____
_____
_____

Name _____ Date _____

Masking Procedures

Objective
Upon completion of this activity sheet, the student should be able to mask a vehicle for refinishing.

ASE Education Foundation Task Correlation

IV.B.8   Mask and protect other areas that will not be refinished. **(HP-I)**

IV.B.9   Demonstrate different masking techniques (recess/back masking, foam door type, etc.). **(HP-G)**

IV.B.16   Remove dust from area to be refinished, including cracks or moldings of adjacent areas. **(HP-I)**

We Support

ASE | Education Foundation

Tools and Equipment
Four-door vehicle
Paper machine with 6-inch, 12-inch, and 18-inch paper
¾-inch and 2-inch tape and other various sizes as necessary
Tire covers
Plastic sheeting car cover
Smooth transition tape
Foam masking tape
Liquid mask
Liquid masking applicator gun

Safety Equipment
Safety glasses or goggles

Introduction
Because paint prep is 95 percent of the paint job, quality masking is one of the most important jobs of the paint process. There is no substitute for taking the necessary time to make sure that each part of the vehicle is properly masked so nothing is painted that is not supposed to be and nothing is taped over that should not be. There is also no substitute for quality material when masking. Good tape will prevent bleed through, peeling, and stuck tape after the job is complete.

This procedure is better understood visually. The required videos listed next will help you familiarize yourself with the materials, masking techniques, and procedures. After viewing the videos, you will be asked to duplicate the masking techniques.

Vehicle Description

Year _____ Make _____ Model _____

Paint code _____ Location of paint code _____

Procedure

1. Go to YouTube, and search the following videos. Watch and study them thoroughly. ☐
    Ask your instructor if there are any additional videos they would recommend. Be sure
    all videos you watch are from a reputable source. The 3M videos listed will give you
    an idea of the proper techniques.
    a.   "3M Soft Edge Foam Masking Tape Plus 06293"—3M Collision Repair Channel
    b.   "Proper Application Techniques for 3M Soft Edge Foam Masking Tape"—3M
         Collision Repair Channel
    c.   "Masking with 3M Overspray Protective Sheeting"—3M Collision Repair Channel
    d.   "Smooth Transition Tape"—3M Collision Repair Channel
    e.   "Proper Process of Masking a Jamb Prior to Paint"—3M Collision Repair Channel
    f.   "Car Painting: How to Mask a Car for Paint like a Pro!"—Paint Society Channel
         (This is a good one to watch.)

2. Now practice on a vehicle in the shop that your instructor has for you. These next
    instructions might make more sense after watching the videos. If you did not watch
    the videos, it will be evident to your instructor. **Do not skip them.**

3. Sprayed paint will find its way onto every surface that is not covered. A quality masking ☐
    job will not only protect the vehicle but also will not have any folds or crevices that may
    trap dirt. In this exercise, you will mask off the side of a four-door vehicle to simulate
    the repainting of an entire side including the rocker panel. Wash the vehicle and blow
    it off using high pressure.
    a.   Why is it important to use high pressure to blow off all the moldings and crevices
         on the vehicle? _____

    _____

4. Begin masking by opening the doors, hood, and deck lid. Start with the ¾-inch tape at ☐
    the front of the rocker panel. Apply a continuous strip all the way to the dog leg. Apply
    pressure to the inside edge of the tape only.
    a.   What can happen if an inexpensive tape is used? _____

    _____

5. Roll the outside edge of the tape back. This is called back taping or rolling the tape. ☐
    It will allow a soft paint edge. Place the 6-inch paper on the tape. You may use foam
    masking tape or smooth transition tape in this procedure where applicable.
    a.   Why is back taping important when you are going to remove the masking tape? ____

    _____

    _____

    b.   What does using the 6-inch paper prevent? _____

    _____

6. Mask off the center pillar by wrapping the paper around the upper portion. Use the ☐
    6-inch paper to mask the rear inner edge of both doors. On the rear door, also mask
    the front inner edge. Do not allow the tape to wrap around the door edge.
    a.   Why is it important to mask the inner edges? _____

    _____

    b.   What will happen if the tape gets on the door edge? _____

    _____

7. Apply the 6-inch paper to the inner edge of the windshield pillar, the inner edge of the
   roof drip rail, and the inner edge of the sail panel. Also mask off the inner edges of the
   window frames, but not the weatherstrip. You may use foam masking tape or smooth
   transition tape in this procedure where applicable.   ☐

   a.   Why is the weatherstrip not taped? _____

   _____

   b.   If the weatherstrip must be masked, wiping it with wax and grease remover usually
        helps with adhesion.

8. Close all the doors, and make sure the paper stays inside the doors. Mask the windows
   by back taping the window opening pinch weld lip. Use an 18-inch paper (or smaller if
   necessary) to attach to the back tape. Once fully taped and sealed down smooth, tape
   the rear window in the same manner.   ☐

   a.   What is the difference in this versus nonpolycoated paper? _____

   _____

9. Apply the 6-inch paper to the underhood fender flange.   ☐

   a.   What will this prevent? _____

   _____

10. Use the 2-inch tape on the hood flange. Close the hood and tape the 18-inch paper
    down the entire hood edge. Mask the windshield at the edge. The neater you can mask,
    the better. Any loose crevice will hold dirt and let overspray buildup and will blow off
    onto your freshly painted panel if you are not careful.   ☐

11. Mask the inner channel of the deck lid with the 6-inch paper. Next, mask the taillight
    openings. Close the lid and apply the paper to the entire deck lid side and the edge
    of the rear window. You may use foam or smooth transition tape for this procedure
    where applicable as well.   ☐

12. Mask the roof at the edge in the same manner as the hood and the deck lid. Plastic
    sheeting can be applied to the remainder of the vehicle.   ☐

    a.   If you are using plastic sheeting, why can you not use it all the way up to the tape
         instead of the paper? _____

    _____

13. Mask the door handle and lock openings by stuffing them with foam or regular masking
    tape. Foam does expand better in some handle openings. Some you can back tape
    easier. They are all slightly different.   ☐

    a.   What method did you use to mask these openings? _____

    _____

    b.   Why would you want to remove the handle and lock when you could just tape them
         off on the car? _____

    _____

14. Wheelhouses can be sprayed with liquid mask.   ☐

    a.   How is liquid mask applied? _____

    _____

    b.   Once the job is complete, how is the liquid mask removed? _____

    _____

Task Completed

15. Tire covers should be blown off and placed over the tires.   ☐

    a. What is the reason for blowing off the tire covers? _____

    _____

16. Wheel openings should be back taped, and paper should be attached to the back tape   ☐
    and draped over the wheel covers. Or, you can back tape, then attach plastic sheeting
    to the back tape, then wrap the sheeting around the wheel, and tuck it under it to seal it
    down tight.

    a. Which method worked best for you? _____

    _____

    b. Explain why. _____

    _____

    _____

INSTRUCTOR'S COMMENTS _____

_____

_____

Name _____ Date _____

Surface Scratch Preparation

Objective

After completion of this activity sheet, the student should be able to properly repair any scratch.

ASE Education Foundation Task Correlation

IV.B.2   Soap and water wash entire vehicle; use appropriate cleaner to remove contaminants. **(HP-I)**

IV.B.3   Inspect and identify type of finish, surface condition, and film thickness; develop and document a plan for refinishing using a total product system. **(HP-G)**

IV.B.4   Remove paint finish as needed. **(HP-I)**

IV.B.5   Dry or wet sand areas to be refinished. **(HP-I)**

IV.B.6   Featheredge areas to be refinished. **(HP-I)**

IV.B.7   Apply suitable metal treatment or primer in accordance with total product systems. **(HP-I)**

IV.B.10   Mix primer, primer-surfacer, and primer-sealer. **(HP-I)**

IV.B.11   Identify a complimentary color or shade of undercoat to improve coverage. **(HP-G)**

IV.B.12   Apply primer onto surface of repaired area. **(HP-I)**

IV.B.14   Block sand area to which primer-surfacer has been applied. **(HP-I)**

IV.B.16   Remove dust from area to be refinished, including cracks or moldings of adjacent areas. **(HP-I)**

IV.B.17   Clean area to be refinished using a final cleaning solution. **(HP-I)**

We Support

ASE | Education Foundation

Tools and Equipment

Any panel
Guide coat
Appropriate spray gun
Urethane primer
#80, #180, #320, #400, and #600 grit paper
Appropriate sanding blocks
Wax and grease remover
Self-etching primer
DA sander

Safety Equipment

Safety glasses or goggles
Dust respirator
Work gloves
Supplied air respirator
Impervious gloves
Paint suit

Introduction

Surface prep is one of the most important aspects of a quality paint job. If the surface is not prepped properly, the new paint will not adhere. Almost all shops give lifetime warranties on their paint, so proper preparation is vital.

Gun Description

Note the following data on the specific spray gun you are working with during this job.

Brand _____

Type _____

Vehicle Description

Year _____ Make _____ Model _____

VIN _____ Paint code _____

Location of paint code _____

Procedure

Task Completed

1. Use an awl to put a 12-inch scratch in the panel. The scratch must be etched into the metal.   ☐

   a. Why is it important to etch the scratch into the metal? _____

   _____

2. Clean the fender with the appropriate soap and water. Dry off the surface. Use a wax and grease remover. Apply the wax and grease remover with a spray bottle and wipe off before it dries.   ☐

   a. What type of soap did you wash the panel with? _____

   b. Why is the type of soap important? _____

   _____

   c. What is the purpose of the wax and grease remover? _____

   _____

   d. Why is it important to wipe off the wax and grease remover before it dries? _____

   _____

3. Featheredge the scratch with #80 grit paper on a DA. Sand about 2 inches from the scratch. Be sure to keep the sander flat. Do not dig into the paint.   ☐

   a. What does DA stand for? _____

   b. What will happen if the paint is not feathered enough? _____

   _____

4. Sand over the #80 grit scratches with #180 or #220 grit paper. Work out from the scratch. Continue sanding about ½ inch beyond the #80 grit scratches with #180 or #220 grit paper. Continue sanding about ½ inch beyond where the #80 grit stopped.   ☐

   a. What is the reason for sanding with the #180 grit? _____

   _____

5. Final sand with #320 grit paper. Sand out from the scratch about ½ inch beyond where the #180 grit stopped. There should be at least 1 inch of each paint layer showing.   ☐

   a. What does this step prevent? _____

   _____

6. Clean off the sanding dust. Mix and spray two coats of self-etching primer.   ☐

   a. Describe the cleaning steps you performed. _____

   _____

   b. Is this primer sanded? _____

**Task
Completed**

7. When the self-etching primer has flashed, spray a urethane primer.   ☐
   a.   How is the urethane primer different from the self-etching primer? _____
   _____

8. Apply a guide coat over the primer.   ☐
   a.   What is the purpose of a guide coat? _____
   _____

9. Block sand the urethane primer with #320 grit, and then wet sand with #400 grit   ☐
   on a block.

10. Finish sanding the urethane primer with wet #600 grit paper.   ☐
   a.   If a panel can be painted after #400 grit sanding, what is the purpose of sanding
   with #600 grit? _____
   _____

INSTRUCTOR'S COMMENTS _____

_____

_____

Review Questions

Name _____ Date _____ Instructor Review _____

1. The paint job is only as good as the repair and preparation job beneath it.
   A. True
   B. False

2. Technician A says that a rubber eraser wheel should be used to remove two-sided tape adhesive left behind from emblems and moldings. Technician B says a rubber eraser wheel cannot be used for pinstripe tape. Who is correct?
   A. Technician A
   B. Technician B
   C. Both Technician A and Technician B
   D. Neither Technician A nor Technician B

3. Open coat sandpaper is used for wet sanding.
   A. True
   B. False

4. Technician A uses paint stripper to remove rust. Technician B uses a #80 grit disc on a DA sander to sand body filler. Who is correct?
   A. Technician A
   B. Technician B
   C. Both Technician A and Technician B
   D. Neither Technician A nor Technician B

5. Wet sanding produces a smoother surface than dry sanding.
   A. True
   B. False

6. A(n) _____ is a light coat of contrasting color intended to show defects.

7. A "bull's-eye" results from improper featheredging.
   A. True
   B. False

8. Technician A uses a spray bottle to apply a wax and grease remover. Technician B uses a rag to apply a wax and grease remover. Who is correct?
   A. Technician A
   B. Technician B
   C. Both Technician A and Technician B
   D. Neither Technician A nor Technician B

9. Technician A wears gloves when rubbing bare metal. Technician B believes that cleaning the bare metal with a wax and grease remover will eliminate oil from the hands. Who is correct?
   A. Technician A
   B. Technician B
   C. Both Technician A and Technician B
   D. Neither Technician A nor Technician B

10. Technician A says that overmasked areas mean that the painter will most of the time have to repaint that panel of the vehicle that should have been painted in the first place. On the other hand, undermasked areas must be cleaned with solvent, clay bar, or polishing compound to remove overspray. Who is correct?
    A. Technician A
    B. Technician B
    C. Both Technician A and Technician B
    D. Neither Technician A nor Technician B

Refinishing Procedures

Name _____ Date _____ Instructor Review _____

Paint Mixing

DUPONT **CHROMABASE®**

CONTAINS: butyl acetate, 123-86-4; acetone, 67-64-1; methyl ethyl ketone, 78-93-3; toluene, 108-88-3; isopropyl alcohol, 67-63-0; ethyl acetate, 141-78-6; propylene glycol monomethyl ether acetate, 108-65-6; xylene, 1330-20-7; ethylene-vinyl acetate resin; cellulosic resin; acrylic polymer.
(For further information refer to Material Safety Data Sheet)

IMPORTANT: May be mixed with other components. Mixture will have hazards of both components. Before opening the packages, read all warning labels. Follow all precautions.

ATTENTION: Contains material which can cause cancer.

NOTICE: Repeated and prolonged overexposure to solvents may lead to permanent brain and nervous system damage. Eye watering, headaches, nausea, dizziness and loss of coordination are signs that solvent levels are too high. Intentional misuse by deliberately concentrating and inhaling the contents may be harmful or fatal.

Do not breathe vapor or spray mist. Do not get in eyes or on skin.

WEAR A PROPERLY FITTED VAPOR/PARTICULATE RESPIRATOR NIOSH/MSHA for use with paints (TC-23C), eye protection, gloves and protective clothing during application and until all vapor and spray mist are exhausted. In confined spaces, or in situations where continuous spray operations are typical, or if proper respirator fit is not possible, wear a positive-pressure, supplied-air respirator (NIOSH/MSHA TC-19C). In all cases, follow respirator manufacturers's directions for respirator use. Do not permit anyone without protection in the painting area.

Keep away from heat, sparks and flame. VAPOR MAY IGNITE EXPLOSIVELY. Vapor may spread long distances. Prevent build-up of vapor. Extinguish all pilot lights and turn off heaters, non-explosion proof electrical equipment and other sources of ignition during and after use and until all vapor is gone. Do not transfer contents to bottles or other unlabeled containers. Close container after each use. Use only with adequate ventilation.

FIRST AID: If affected by inhalation of vapor or spray mist, remove to fresh air. In case of eye contact, flush immediately with plenty of water for at least 15 minutes and call a physician; for skin, wash thoroughly with soap and water. If swallowed, CALL A PHYSICIAN IMMEDIATELY. DO NOT induce vomiting.

SPILL/WASTE: Absorb spill and dispose of waste or excess material according to Federal, State and local regulations.

KEEP OUT OF REACH OF CHILDREN   PHOTOCHEMICALLY REACTIVE

DIRECTIONS FOR USE

1 Part   +   1 Part   ⟶   Spray at
Color   Basemaker®   45 psi.

MIXING: To each quart of base color, add one quart of the appropriate Basemaker: 7160S Low Temp, 7175S Mid Temp, 7185S High Temp or 7195S Very High Temp.

PREPARATION: Prepare all surfaces to be repaired using a DuPont undercoat system following recommended procedures. Finish by sanding with 400 grit paper (wet or dry).

SPOT REPAIR/BLENDING: Midcoat entire area to be repaired with one coat of 222S Mid-Coat Adhesion Promoter. Apply base color at 45 psi in 2-3 medium coats to achieve visual hiding. Extend each coat of base color beyond the previous one to achieve a tapered edge keeping within the midcoated area.

PANEL/OVERALL APPLICATION: To insure uniformity, apply 1 coat of Prime 'N Seal™ or VELVASEAL™ per label directions over entire area to be repaired. Apply base color to hiding.
Flash Basecoat 15-30 minutes before clearcoating.

TWO TONING: Flash first base color 30 minutes and then tape. Apply second color, remove tape and flash 30 minutes prior to clearcoating.

FOR VOC REGULATED AREAS: These directions refer to the use of products which may be restricted or require special mixing instructions in VOC regulated areas. Follow mixing and usage recommendations in the VOC Compliant Products Chart for your area.

(See Individual Label Directions For Use Of The Other DuPont Products)

Made in U.S.A   E-R1301   H-07017   3RF-548ABAK-040-0294
E. I. DU PONT DE NEMOURS & CO.(INC.), Wilmington, Delaware 19898
For medical & environmental information: (800) 441-7515

---

1. What is the spray viscosity? _____

2. At what air pressure is this material sprayed? _____

3. What safety equipment must be worn when spraying? _____

4. What basemaker (reducer) is used if the shop temperature is 70 degrees? _____

5. What is the mixing ratio? _____

6. What is the pot life? _____

7. How many coats should be applied? _____

DUPONT   7600 S   CHROMACLEAR®
SUPER PRODUCTIVE

CONTAINS: acrylic polymer; butyl acetate, 123-86-4; methyl ethyl ketone, 78-93-3; methyl isobutyl ketone, 108-10-1;   toluene, 108-88-3; xylene, 1330-20-7. When mixed with activator, also contains aliphatic polyisocyanate resin and hexamethylene diisocyanate monomer; 28182-81-2.
(For further information refer to Material Safety Data Sheet)

IMPORTANT: Must be mixed with other components. Mixture will have hazards of both components. Before opening the packages, read all warning labels. Follow all precautions.

NOTICE: Repeated and prolong overexposure to solvents may lead to permanent brain and nervous system damage. Eye watering, headaches, nausea, dizziness and loss of coordination are signs that the solvent levels are too high. Intentional misuse by deliberately concentrating and inhaling the contents may be harmful or fatal.

Do not breathe vapor or spray mist. Do not get in eyes or on skin.

WEAR A POSITIVE-PRESSURE, SUPPLIED-AIR RESPIRATOR (NIOSH/MSHA TC-19C), EYE PROTECTION, GLOVES AND PROTECTIVE CLOTHING WHILE MIXING ACTIVATOR WITH ENAMEL, DURING APPLICATION AND UNTIL ALL VAPOR AND SPRAY MIST ARE EXHAUSTED. Follow respirator manufacturer's directions for respirator use.

INDIVIDUALS WITH HISTORY OF LUNG OR BREATHING PROBLEMS OR PRIOR REACTION TO ISOCYANATES SHOULD NOT USE OR BE EXPOSED TO THIS PRODUCT. Do not permit anyone without protection in the painting area.

Keep away from heat, sparks and flame. VAPOR MAY CAUSE FLASH FIRE. Do not transfer contents to bottles or other unlabeled containers. Close container after each use. Use only with adequate ventilation.

FIRST AID: If affected by inhalation of vapor or spray mist, remove to fresh air. If breathing difficulty persists, or occurs later, consult a physician. In case of eye contact, flush immediately with plenty of water for at least 15 minutes and call a physician; for skin, wash thoroughly with soap and water. If swallowed, call a physician immediately and have label information available. DO NOT induce vomiting.

SPILL/WASTE: Absorb spill and dispose of waste or excess material according to Federal, State and local regulations.

KEEP OUT OF REACH OF CHILDREN

PHOTOCHEMICALLY REACTIVE

DIRECTIONS FOR USE

ACTIVATON: The activation ratio for 7600 S is 4 to 1.   That is, to one quart of 7600 S, add ½ pint (two-4 oz. overcaps) of the appropriate activator-reducer: 7655 S Spot Activator-Reducer; 7675 S Panel Activator-Reducer; or 7695 S Multi Panel Activator-Reducer. Pot life of the activated clear is 3 hours.

APPLICATION: Stir thoroughly and strain. For spot and panel repairs, use 25-35 lbs. air pressure at the gun. For overalls, use 45-55 lbs. air pressure at the gun. Apply 2-3 medium coats. Allow 1 to 5 minutes flash between coats, depending on air flow and temperature.

FORCE DRY: Allow final coat to flash between 5-10 minutes at ambient temperatures before force drying. Force dry at 140°F (60°C) for 30 minutes. Can be polished and delivered after 2 hour cool-down.

EXPRESS DRY: Allow final coat of clear to flash 3-8 minutes at ambient temperatures. Express Dry at 120°F for 15 minutes. Can be polished and delivered after a 4 hour cool-down.

AIR DRY: Can be polished and delivered within 4-6 hours depending on ambient temperatures.

POLISHING: For dirt removal, lightly nib sand with 1500 grit paper and polish with either DuPont 600 S or 1500 S.

FISH EYES: Use up to 1 oz. of 259 S Additive per gallon of 7600 S. DO NOT USE FEE.

CLEAN UP: Clean all equipment immediately after use with DuPont lacquer thinner.

FOR VOC REGULATED AREAS: These directions refer to the use of products which may be restricted or require special mixing instructions in VOC regulated areas. Follow mixing and usage recommendations in the VOC Compliant Products Chart for your area.

Made in U.S.A.   (SEE INDIVIDUAL LABEL DIRECTIONS FOR USE OF THE OTHER DU PONT PRODUCTS)   3RF-266BBES-040-0193

F. I. DU PONT DE NEMOURS & CO. (INC.), Wilmington, Delaware 19898
For medical & environmental information: (800) 441-7515

1. What is the spray viscosity? _____

2. At what air pressure is this material sprayed? _____

3. What safety equipment must be worn when spraying? _____

4. What hardener is used if the shop temperature is 80 degrees? _____

5. What is the mixing ratio? _____

6. What is the pot life? _____

7. How many coats should be applied? _____

Name _____ Date _____

Paint Mixing and Spraying Basecoat/Clearcoat

Objective

Upon completion of this activity sheet, the student should be able to mix two- and three-part paints and to test viscosity.

ASE Education Foundation Task Correlation

IV.C.1	Inspect, clean, and determine condition of spray guns and related equipment (air hoses, regulators, air lines, air source, and spray environment). **(HP-I)**
IV.C.2	Select spray gun setup (fluid needle, nozzle, and cap) for product being applied. **(HP-I)**
IV.C.3	Test and adjust spray gun using fluid, air, and pattern control valves. **(HP-I)**
IV.C.4	Demonstrate an understanding of the operation of pressure spray equipment. **(HP-G)**
IV.D.1	Identify color code by manufacturer's vehicle information label. **(HP-I)**
IV.D.2	Shake, stir, reduce, catalyze/activate, and strain refinish materials. **(HP-I)**
IV.D.3	Apply finish using appropriate spray techniques (gun arc, angle, distance, travel speed, and spray pattern overlap) for the finish being applied. **(HP-I)**
IV.D.6	Apply basecoat/clearcoat for panel blending and panel refinishing. **(HP-I)**
IV.D.7	Apply basecoat/clearcoat for overall refinishing. **(HP-G)**
IV.D.8	Remove nibs or imperfections from basecoat. **(HP-I)**
IV.D.12	Identify and mix paint using a formula. **(HP-I)**
IV.E.4	Identify lifting; correct the cause(s) and the condition. **(HP-G)**
IV.E.5	Identify clouding (mottling and streaking in metallic finishes); correct the cause(s) and the condition. **(HP-I)**
IV.E.9	Identify sags and runs in paint surface; correct the cause(s) and the condition. **(HP-I)**
IV.E.10	Identify sanding marks or sand scratch swelling; correct the cause(s) and the condition. **(HP-I)**
IV.E.12	Identify color difference (off-shade); correct the cause(s) and the condition. **(HP-G)**
IV.E.15	Identify poor adhesion; correct the cause(s) and the condition. **(HP-G)**
IV.E.16	Identify paint cracking (shrinking, splitting, crowsfeet or line-checking, micro-checking, etc.); correct the cause(s) and the condition. **(HP-G)**
IV.E.18	Identify dirt or dust in the paint surface; correct the cause(s) and the condition. **(HP-I)**

We Support
ASE | Education Foundation

Tools and Equipment

Any panel designated by your instructor
Sealer
Basecoat and clearcoat
Appropriate reducers and catalysts (hardeners)
Mixing cups

Safety Equipment

Safety glasses or goggles
Paint respirator
Impervious gloves
Paint suit

Mixing sticks
Paint shaker
Zahn cup
Stopwatch
Graduated mix cup
Tack rag
HVLP or compliant gun
Wax and grease remover
Gun cleaning station

Introduction

With some of the newer factory paints costing hundreds of dollars per quart, it is critical that paint be mixed and sprayed correctly. A lot of shops now have their own mixing stations because it saves money and creates less waste.

Procedure

Task Completed

Paint Mixing

1. Read the mixing instructions on the paint can or paint manual. Some paints are two or   ☐
   three stage. It can be a basecoat/clearcoat (two stage) or a pearl tri-coat (three stage)
   that requires using a tinted mid-coat for accurate color match. This varies from car to
   car depending on the paint code from the manufacturer. Decide on how much paint
   material you will need and what kind you are spraying.

   a. What brand of paint are you mixing? _____

   b. Is this a regular basecoat/clearcoat (two stage) or a tri-coat (three stage) basecoat/
      clearcoat?

      _____

2. Place the basecoat in the paint shaker for 5–10 minutes. If there is no shaker   ☐
   available, open the can and stir vigorously for 5–10 minutes. All solvent primers, sealers,
   and basecoats must be agitated before mixing.

   a. What can happen to metallic colors if they are not properly mixed? _____

      _____

   b. Will not mixing paint properly affect the color when spraying? If so, how? _____

      _____

Prepping and Spraying Basecoat/Clearcoat

3. Wash the panel or car using soap and water, and then use a wax and grease remover   ☐
   to remove any leftover contaminants before you begin the prep procedures.

4. With the addition of waterborne paints, the prep procedures can vary depending on   ☐
   the brand and type of paint you are using. Also, it can vary depending if it is a solid
   color or a metallic. Solid colors you can typically stop at 600 grit at the finest. For
   solvent metallic basecoats and all waterborne basecoats, you need to step it up to at
   least 800 grit to keep scratches to a minimum. Ask your instructor what type of paint
   you are going to use for this procedure. Use a squeegee to remove the excess water as
   you sand so it does not dry on the panel and stain the panel.

   a. What type of basecoat (waterborne or solvent) are you using? And is it a solid
      color or metallic? _____

   b. What can happen if the panel is not sanded thoroughly? _____

      _____

5. Clean the panel again with a wax and grease remover. Use a spray bottle to distribute   ☐
the remover. Wipe it off before the panel dries.

   a. Why is it important to wipe off the solvent before it dries? _____

    _____

6. Mix the basecoat according to the product specifications. Refer to the technical   ☐
data sheet if needed.

   a. What is the mix ratio? _____

   b. If this is a metallic color, what precautions must be taken to spray it? _____

    _____

   c. What can happen if the paint is sprayed on too wet? _____

    _____

7. Blow the panel off and tack rag the entire surface being refinished so that you remove   ☐
any lint particles and dust from the surface.

8. Spray one coat of sealer over the primered area being refinished or the entire panel   ☐
if it is a new panel. Wait 10–15 minutes to flash off before spraying the basecoat. Once
dry, glide the tack rag over the area being refinished again to remove any lint or dust
particles.

   a. What can happen if you do not run the tack rag over the surface before basecoat? _____

    _____

9. Spray one coat of paint as per the paint manufacturer's recommendations over the   ☐
sealer. Allow this coat to flash accordingly depending on the manufacturer's
recommendations.

   a. How long is the flash time? _____

10. Wipe the area being painted again with the tack rag once the flash time has been reached.   ☐
Spray a second coat. Allow proper flash time, and spray a third coat. On metallic colors,
spray a drop coat. This is a dryer coat than the last two coats where you step your gun
back more 3–4 inches from the surface area and spray a coat over the entire area that
was refinished. This will help the metallic pigments lay down more uniform and even.
Ask your instructor for help with this procedure if needed.

11. Clean the basecoat out of the gun making sure to wear protective safety equipment.   ☐
Mix your clearcoat, and load it in the gun.

   a. What is the mix ratio of the clear? _____

12. Before applying the clearcoat, tack the surface again. Be careful to only lightly wipe   ☐
the panel.

   a. What can happen if you wipe too hard with the tack rag on fresh paint? _____

    _____

13. Spray two coats of clear. Most manufacturers recommend two full wet coats.   ☐

   a. What is the proper flash time for the clear? _____

14. Check the panel for any defects such as scratches, lifting, bull's-eyes, and so on.   ☐

   a. What is the drying time before the panel can be wet sanded and buffed or
reassembled?

    _____

15. Clean the clearcoat out of the gun making sure to wear protective safety □
    equipment.

INSTRUCTOR'S COMMENTS _____

_____

_____

Name _____ Date _____

Edging Panels

Objective

Upon completion of this activity sheet, the student should be able to edge parts prior to installation on a vehicle.

ASE Education Foundation Task Correlation

IV.B.3	Inspect and identify type of finish, surface condition, and film thickness; develop and document a plan for refinishing using a total product system. **(HP-G)**
IV.B.4	Remove paint finish as needed. **(HP-I)**
IV.B.5	Dry or wet sand areas to be refinished. **(HP-I)**
IV.B.10	Mix primer, primer-surfacer, and primer-sealer. **(HP-I)**
IV.B.11	Identify a complimentary color or shade of undercoat to improve coverage. **(HP-G)**
IV.B.16	Remove dust from the area to be refinished, including cracks or moldings of adjacent areas. **(HP-I)**
IV.B.17	Clean area to be refinished using a final cleaning solution. **(HP-I)**
IV.B.18	Remove, with a tack rag, any dust or lint particles from the area to be refinished. **(HP-I)**
IV.B.19	Apply suitable primer-sealer to the area being refinished. **(HP-I)**
IV.B.20	Scuff sand to remove nibs or imperfections from a sealer. **(HP-I)**
IV.C.1	Inspect, clean, and determine condition of spray guns and related equipment (air hoses, regulators, air lines, air source, and spray environment). **(HP-I)**
IV.C.2	Select spray gun setup (fluid needle, nozzle, and cap) for product being applied. **(HP-I)**
IV.C.3	Test and adjust spray gun using fluid, air, and pattern control valves. **(HP-I)**
IV.C.4	Demonstrate an understanding of the operation of pressure spray equipment. **(HP-G)**
IV.D.1	Identify color code by manufacturer's vehicle information label. **(HP-I)**
IV.D.2	Shake, stir, reduce, catalyze/activate, and strain refinish materials. **(HP-I)**
IV.D.3	Apply finish using appropriate spray techniques (gun arc, angle, distance, travel speed, and spray pattern overlap) for the finish being applied. **(HP-I)**
IV.D.6	Apply basecoat/clearcoat for panel blending and panel refinishing. **(HP-I)**

We Support

ASE | Education Foundation

Tools and Equipment

Any new panel
Sealer
Basecoat
Clearcoat
HVLP, RP, or an EPA-compliant paint gun
Scuff pad
Wax and grease remover
Tack rag

Safety Equipment

Safety glasses or goggles
Appropriate respirator
Impervious gloves
Paint suit

Introduction

Many times the difference between an unprofessional paint job and an excellent one is in the prep. If the time is not taken to properly edge a new panel, the signs of this neglect will be very obvious. This will reduce the trade-in value of the car.

Vehicle Description

Year _____ Make _____ Model _____

VIN _____

Procedure

Task Completed

1. Check the new part for dents. Make sure it is the correct part for the year, side, and so on. ☐
   a. If there is any damage, how do you determine whether to fix the panel or return it?

   _____

2. Check the original part to see what areas are to be painted. You must apply color on ☐
   the new part to match what is painted on the original part.

3. Clean the part with a wax and grease remover. ☐

4. Sand all the areas to be painted with a coarse scuff pad or #400 grit paper. Be careful ☐
   because many new parts have very sharp edges. Remove all shiny spots, and then
   remove all sanding residue.

5. Mount the panel on a paint stand, remove all sanding residue by blowing it off with an ☐
   airblower, and then pre-clean the area again with wax and grease remover.

   a. If the panel has a factory primer on it, why must it be degreased, scuffed, and
      sealed?

   _____

6. Mix and spray sealer or epoxy primer on the entire sanded area. ☐

   a. Are there isocyanates in the sealer/basecoat/clearcoat?_____

   b. If so, what precautions must be taken? ?_____

   _____

7. Mix and spray basecoat on the areas that cannot be sprayed after the part is mounted ☐
   on the vehicle.

   a. What brand paint did you use, and how did you mix it? _____

   _____

   b. Why is it critical for this step to be done now and not after the part is bolted to the
      vehicle?

   _____

   _____

8. On a fender, this would be the large flange under the hood and the lip visible when the ☐
   door is opened. You should spray the front flange, the wheel well lip, and the bottom
   areas. On a door, spray the entire inner and outer window channel. On a hood or deck
   lid, spray the inside and edge panel. Allow the proper flash time between coats.

   a. What is the proper flash time? _____

9. Some paint manufacturers make a single part edging clear. It is meant to be sprayed ☐
   on areas that will not be in sunlight. Simply load it into the gun and spray it on.

   a. Why can this clear not be used on the outside of the panel? _____

   _____

**Task
Completed**

10. Mix and spray clearcoat over the basecoat.   ☐

    a. What is the brand of clear you are spraying? ‗‗‗‗‗‗‗

    ‗‗‗‗‗‗‗‗‗‗‗‗‗‗‗‗‗‗‗‗‗‗‗‗‗‗‗‗‗‗‗‗‗

    b. What is the mix ratio? ‗‗‗‗‗‗‗‗‗‗‗‗‗‗‗‗‗‗‗

INSTRUCTOR'S COMMENTS ‗‗‗‗‗‗‗‗‗‗‗‗‗‗‗‗‗‗‗‗‗‗‗‗‗‗‗

‗‗‗

‗‗‗

Name _____ Date _____

Paint and Blend Procedures

Objective

Upon completion of this activity sheet, the student should be able to correctly paint and blend a fender using basecoat/clearcoat.

ASE Education Foundation Task Correlation

IV.B.1   Inspect, remove, store, protect, and replace exterior trim and components necessary for proper surface preparation. **(HP-I)**

IV.B.2   Soap and water wash entire vehicle; use appropriate cleaner to remove contaminants. **(HP-I)**

IV.B.3   Inspect and identify type of finish, surface condition, and film thickness; develop and document a plan for refinishing using a total product system. **(HP-G)**

IV.B.4   Remove paint finish as needed. **(HP-I)**

IV.B.5   Dry or wet sand areas to be refinished. **(HP-I)**

IV.B.6   Featheredge areas to be refinished. **(HP-I)**

IV.B.7   Apply suitable metal treatment or primer in accordance with total product systems. **(HP-I)**

IV.B.8   Mask and protect other areas that will not be refinished. **(HP-I)**

IV.B.9   Demonstrate different masking techniques (recess/back masking, foam door type, etc.). **(HP-G)**

IV.B.10   Mix primer, primer-surfacer, and primer-sealer. **(HP-I)**

IV.B.11   Identify a complimentary color or shade of undercoat to improve coverage. **(HP-G)**

IV.B.16   Remove dust from the area to be refinished, including cracks or moldings of adjacent areas. **(HP-I)**

IV.B.17   Clean area to be refinished using a final cleaning solution. **(HP-I)**

IV.B.18   Remove, with a tack rag, any dust or lint particles from the area to be refinished. **(HP-I)**

IV.B.19   Apply suitable primer-sealer to the area being refinished. **(HP-I)**

IV.B.20   Scuff sand to remove nibs or imperfections from a sealer. **(HP-I)**

IV.B.23   Prepare adjacent panels for blending. **(HP-I)**

IV.B.25   Identify metal parts to be refinished; determine the materials needed, preparation, and refinishing procedures. **(HP-I)**

IV.C.1   Inspect, clean, and determine condition of spray guns and related equipment (air hoses, regulators, air lines, air source, and spray environment). **(HP-I)**

IV.C.2   Select spray gun setup (fluid needle, nozzle, and cap) for product being applied. **(HP-I)**

IV.C.3   Test and adjust spray gun using fluid, air, and pattern control valves. **(HP-I)**

IV.C.4   Demonstrate an understanding of the operation of pressure spray equipment. **(HP-G)**

IV.D.2   Shake, stir, reduce, catalyze/activate, and strain refinish materials. **(HP-I)**

IV.D.3   Apply finish using appropriate spray techniques (gun arc, angle, distance, travel speed, and spray pattern overlap) for the finish being applied. **(HP-I)**

IV.D.6   Apply basecoat/clearcoat for panel blending and panel refinishing. **(HP-I)**

We Support

ASE | **Education Foundation**

Tools and Equipment

Vehicle with fender (or any panel) to replace
Wax and grease remover
#600 and #800 grit paper
Masking tape and paper
Basecoat
Clearcoat
HVLP gun
Tack rag

Safety Equipment

Safety glasses or goggles
Appropriate respirator
Impervious gloves
Paint suit

Introduction

In early automotive painting, there were only a few solid colors available for painting vehicles. Today, there are hundreds of colors and types of paint to choose from. When vehicles are damaged and need to be repainted, it is vital that there be no visible sign of repair.

Vehicle Description

Year _____ Make _____ Model _____

VIN _____

Procedure

Task Completed

1. Wash the car and de-wax the area to be refinished.   ☐

   a. What is the importance of using a detergent-type soap (Dawn, Joy, Gain, etc.) before you paint a car? _____

   b. What is the importance of using wax and grease remover? _____
   _____

2. Remove all moldings, nameplates, and other trim from the adjacent panels to the panel being replaced.   ☐

3. If the fender has been replaced, assume that it has been edged. Sand the new panel that was replaced with #600 grit paper for solvent paint and #800 grit paper for waterborne paint.   ☐

4. Take a scuff pad around the edges of the panel and the high crowns first. Gray for metallics and pearl paints. Red for solid colors. Be very, very careful not to sand through the paint. The edges are the easiest to cut through! They just need to be dull and no shiny left once scuffed.   ☐

5. Sand the clearcoat on the adjacent panels surrounding the new panel with #800 grit paper. This can be done with wet sanding paper or dry sanding paper on a DA and a foam backing pad. Sand the orange peel out of the clearcoat. Be very careful not to sand through the paint. Blow the dust and/or water from all the crevices and gaps. Wipe with a mild wax and grease remover.   ☐

   a. Why do you sand the panels with the finer grit (#800) for waterborne paint and not for solvent? _____
   _____

   b. Why is it important to sand out the orange peel? _____

   c. What can happen if a wax and grease remover is not used?
   _____

6. Mask off the vehicle, blow it off, and tack. □

7. Spray one coat of sealer per the manufacturer's recommendations. Allow to flash per the recommendations, and then tack the surface. □

8. Spray one coat of base on the new panel only. Allow this to flash accordingly. Tack off the adjacent panels surrounding the new panel. □

9. Spray the second coat. End the second coat 3–4 inches on the surrounding panels. Allow the coat to flash accordingly. □

10. Reduce the color by 50 percent, and respray another 3–4 inches into the door and hood. Reduce again by 50 percent and respray, repeating the previous step. □

11. Tack off the entire area that is being refinished including the paper surrounding the panels. Mix and spray the first coat of clear over the entire area. Allow the coat to flash accordingly. □

   a. What is the mix ratio for the clear? _____

12. Spray the second coat of clear over the entire area. Make sure that it is a full wet coat. □

   a. What is the cure time? _____

13. Check for any surface defects. If there are any, refer to the proper job sheet for removal of certain defects. □

INSTRUCTOR'S COMMENTS _____

_____

_____

Name _____ Date _____

Blending a Sail Panel

Objective

Upon completion of this activity sheet, the student should be able to melt the refinish clearcoat into the original clearcoat. This is the usual procedure when the quarter panel is replaced and the paint and clear are blended into the sail panel.

ASE Education Foundation Task Correlation

IV.B.1	Inspect, remove, store, protect, and replace exterior trim and components necessary for proper surface preparation. **(HP-I)**
IV.B.2	Soap and water wash entire vehicle; use appropriate cleaner to remove contaminants. **(HP-I)**
IV.B.3	Inspect and identify type of finish, surface condition, and film thickness; develop and document a plan for refinishing using a total product system. **(HP-G)**
IV.B.4	Remove paint finish as needed. **(HP-I)**
IV.B.5	Dry or wet sand areas to be refinished. **(HP-I)**
IV.B.6	Featheredge areas to be refinished. **(HP-I)**
IV.B.7	Apply suitable metal treatment or primer in accordance with total product systems. **(HP-I)**
IV.B.8	Mask and protect other areas that will not be refinished. **(HP-I)**
IV.B.9	Demonstrate different masking techniques (recess/back masking, foam door type, etc.). **(HP-G)**
IV.B.10	Mix primer, primer-surfacer, and primer-sealer. **(HP-I)**
IV.B.11	Identify a complimentary color or shade of undercoat to improve coverage. **(HP-G)**
IV.B.16	Remove dust from the area to be refinished, including cracks or moldings of adjacent areas. **(HP-I)**
IV.B.17	Clean area to be refinished using a final cleaning solution. **(HP-I)**
IV.B.18	Remove, with a tack rag, any dust or lint particles from the area to be refinished. **(HP-I)**
IV.B.19	Apply suitable primer-sealer to the area being refinished. **(HP-I)**
IV.B.20	Scuff sand to remove nibs or imperfections from a sealer. **(HP-I)**
IV.B.23	Prepare adjacent panels for blending. **(HP-I)**
IV.B.24	Identify the types of rigid, semi-rigid, or flexible plastic parts to be refinished; determine the materials needed, preparation, and refinishing procedures. **(HP-I)**
IV.B.25	Identify metal parts to be refinished; determine the materials needed, preparation, and refinishing procedures. **(HP-I)**
IV.C.1	Inspect, clean, and determine condition of spray guns and related equipment (air hoses, regulators, air lines, air source, and spray environment). **(HP-I)**
IV.C.2	Select spray gun setup (fluid needle, nozzle, and cap) for product being applied. **(HP-I)**
IV.C.3	Test and adjust spray gun using fluid, air, and pattern control valves. **(HP-I)**
IV.D.1	Identify color code by manufacturer's vehicle information label. **(HP-I)**
IV.D.2	Shake, stir, reduce, catalyze/activate, and strain refinish materials. **(HP-I)**
IV.D.3	Apply finish using appropriate spray techniques (gun arc, angle, distance, travel speed, and spray pattern overlap) for the finish being applied. **(HP-I)**

IV.D.6   Apply basecoat/clearcoat for panel blending and panel refinishing. **(HP-I)**

IV.D.8   Remove nibs or imperfections from basecoat. **(HP-I)**

IV.D.10   Refinish plastic parts. **(HP-I)**

IV.D.12   Identify and mix paint using a formula. **(HP-I)**

IV.D.14   Tint color using formula to achieve a blendable match. **(HP-I)**

IV.D.15   Identify alternative color formula to achieve a blendable match. **(HP-I)**

We Support

ASE | **Education Foundation**

Tools and Equipment

Any vehicle
Basecoat
Clearcoat
Masking tape and paper
HVLP gun
Adhesion promoter
Rubbing compound
Wax and grease remover
Tack rag
Blending clear

Safety Equipment

Safety glasses or goggles
Supplied air respirator
Impervious gloves
Paint suit

Introduction

Blending paint into a sail panel used to be a common practice in some paint shops. Most shops now choose to refinish the entire panel so there is no burn in area that could start to peel back later on. This job sheet will walk you through the steps on how to properly complete a burn in or blending a sail panel.

Vehicle Description

Year _____ Make _____ Model _____

VIN _____

Procedure

Task Completed

1. Obtain a vehicle in which the quarter panel can be painted and the sail panel blended. It is easier to melt into a narrow panel. If the panel is wider than 18 inches, this technique is quite difficult. Do not try to blend a large area such as the roof or hood.   ☐

   a. Why is it easier to blend into a narrow panel? _____

   _____

   b. Why is it not advisable to blend into the roof or hood? _____

   _____

2. Clean and de-wax the vehicle.   ☐

   a. How is the vehicle de-waxed? _____

3. If the quarter panel has been replaced, block and sand the primer as needed. You may need to blend the deck lid, the bumper, and the door. Use job sheet 28-3 for this procedure. Sand the area to be cleared with wet #1000 grit paper.   ☐

   a. What sandpaper is used to sand the primer? _____

b. Why is wet #1000 grit used instead of #400 grit on the area to be blended? _____

_____

4. After the surfaces have been prepped, blow off all the cracks and crevices. Use a wax ☐
and grease remover to clean the areas to be painted.

   a. What is the purpose of using a wax and grease remover?

5. Check the paint manufacturer's recommendations for adhesion promoter. In some cases, ☐
it is required to bind the refinish paint to the original clearcoat. If adhesion promoter is
required, follow the instructions. It has a tendency to run, so be careful. Spray it over the
rubbed area. Some paint manufacturers recommend spraying it over the entire refinish
area.

   a. What is the purpose of the adhesion promoter? _____

   b. What can happen if adhesion promoter is not used? _____

6. Mix and spray sealer over the sanded primer area that was repaired. ☐

7. Now you must find a blendable match between the different alternates of the paint code, ☐
and mix the best matching basecoat you can find for this code.

   a. What was the variant you used for this code? _____

_____

8. Tint the color if you could not find an achievable blend for this color. It is best to stay ☐
within the colors that are used to make up the paint code as much as possible. Most
of the time you can achieve a blendable color staying within the formula versus going
outside the formula and using other tints.

   a. List here what colors and how much of each you used to tint this color. _____

_____

   b. How did the color match after you tinted and sprayed a test card of this color? ____

_____

9. Now spray the first coat of base on the sealer only. Allow the coat to flash, and tack the ☐
panel once dried completely with a tack cloth. _____

   a. What is the purpose of using a tack cloth?

_____

   b. What can happen if you tack the paint too soon? _____

_____

10. Fan the spray further with each pass of the gun, painting past the original work area until ☐
blended.

11. Clean the spray gun. Most shops are now equipped with a gun cleaner for safe and ☐
easier cleaning.

12. Now mix the clearcoat.

   a. What brand clearcoat are you using, and what is the mix ratio? _____

_____

13. Load the gun with clear. Mix up some clear blender in a second paint gun. This is called ☐
the two-gun method.

   a. What is the mix ratio for the blending of clear? _____

   b. Is it sprayed any differently than the overall clear? If so, how? _____

_____

14. The first coat of clear should never cover all of the paint and end in the rubbed area. ☐
    As soon as you spray this first coat, change guns and spray the blending clear over the
    dry edge of the clear coat.

    a. What does the blending clear do? _____

15. Spray the last coat of clear out about 1 inch from the edge of the first coat. Again, switch ☐
    guns and melt the dry clear edge with the blending clear.

INSTRUCTOR'S COMMENTS _____

_____

_____

Review Questions

Name _____ Date _____ Instructor Review _____

1. Technician A plans to blend all repairs. Technician B checks the color before planning to blend. Who is correct?
   A. Technician A
   B. Technician B
   C. Both Technician A and Technician B
   D. Neither Technician A nor Technician B

2. Technician A states that surface prep is the most important factor in a good paint job. Technician B believes that the topcoat is the most important factor. Who is correct?
   A. Technician A
   B. Technician B
   C. Both Technician A and Technician B
   D. Neither Technician A nor Technician B

3. Technician A says that spot painting is the easiest type of repair. Technician B believes that panel painting is the easiest type of repair. Who is correct?
   A. Technician A
   B. Technician B
   C. Both Technician A and Technician B
   D. Neither Technician A nor Technician B

4. ____ should be used when refinishing plastic parts to promote proper bonding.

5. Primer surfaces should be sprayed as heavy coats with little flash time between coats.
   A. True
   B. False

6. Technician A says that sealer is used to provide adhesion between the paint and the substrate. Technician B believes that sealer prevents bleed-through. Who is correct?
   A. Technician A
   B. Technician B
   C. Both Technician A and Technician B
   D. Neither Technician A nor Technician B

7. Some manufacturers make a clearcoat that dries to the touch in 20 minutes.
   A. True
   B. False

8. In a downdraft booth, a respirator is not needed when spraying any refinish materials.
   A. True
   B. False

9. Technician A states that a slow-drying reducer is used in hot weather. Technician B believes that a slow-drying reducer is used in cold weather. Who is correct?
   A. Technician A
   B. Technician B
   C. Both Technician A and Technician B
   D. Neither Technician A nor Technician B

10. Technician A says that proper airflow is the most important factor of drying waterborne basecoat. Technician B says that temperature is the most important factor of drying waterborne basecoat. Who is correct?
    A. Technician A
    B. Technician B
    C. Both Technician A and Technician B
    D. Neither Technician A nor Technician B

Color Matching and Custom Painting

Name _____ Date _____ Instructor Review _____

Color Matching

Define the following terms explaining color matching.

1. White light _____

2. Color spectrum _____

3. Incandescent light _____

4. Fluorescent light _____

5. Lumen rating _____

6. Value _____

7. Hue _____

8. Chroma _____

9. Metamerism _____

10. Color directory _____

11. Basecoat patch _____

12. Intermix system _____

13. Variance chips _____

14. Flop _____

15. Spray-out panel _____

16. Let-down panel _____

17. Halo effect _____

18. Mica _____

19. Tinting _____

20. Solids _____

21. Metallics _____

22. Pearls _____

Name _____ Date _____

Tri-Coat Let-Down Panel

Objective

Upon completion of this activity sheet, the student should be able to construct a let-down panel to check the color match on a tri-coat finished vehicle.

ASE Education Foundation Task Correlation

IV.B.3	Inspect and identify type of finish, surface condition, and film thickness; develop and document a plan for refinishing using a total product system. **(HP-G)**
IV.B.5	Dry or wet sand areas to be refinished. **(HP-I)**
IV.B.6	Featheredge areas to be refinished. **(HP-I)**
IV.B.11	Identify a complimentary color or shade of undercoat to improve coverage. **(HP-G)**
IV.B.16	Remove dust from the area to be refinished, including cracks or moldings of adjacent areas. **(HP-I)**
IV.B.17	Clean area to be refinished using a final cleaning solution. **(HP-I)**
IV.B.18	Remove, with a tack rag, any dust or lint particles from the area to be refinished. **(HP-I)**
IV.B.19	Apply suitable primer-sealer to the area being refinished. **(HP-I)**
IV.B.20	Scuff sand to remove nibs or imperfections from a sealer. **(HP-I)**
IV.C.1	Inspect, clean, and determine condition of spray guns and related equipment (air hoses, regulators, air lines, air source, and spray environment). **(HP-I)**
IV.C.2	Select spray gun setup (fluid needle, nozzle, and cap) for product being applied. **(HP-I)**
IV.C.3	Test and adjust spray gun using fluid, air, and pattern control valves. **(HP-I)**
IV.C.4	Demonstrate an understanding of the operation of pressure spray equipment. **(HP-G)**
IV.D.1	Identify color code by manufacturer's vehicle information label. **(HP-I)**
IV.D.2	Shake, stir, reduce, catalyze/activate, and strain refinish materials. **(HP-I)**
IV.D.3	Apply finish using appropriate spray techniques (gun arc, angle, distance, travel speed, and spray pattern overlap) for the finish being applied. **(HP-I)**
IV.D.4	Apply selected product on test or let-down panel; check for color match. **(HP-I)**
IV.D.11	Apply multistage coats for panel blending and overall refinishing. **(HP-G)**

We Support

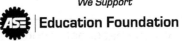
Education Foundation

Tools and Equipment

3-foot × 3-foot piece of sheet metal or any test panel
Self-etching primer
Fine line tape
Masking paper
HVLP, RP, or any EPA-compliant spray gun
White urethane sealer
Basecoat
Pearlcoat or tri-coat

Safety Equipment

Safety glasses or goggles
Supplied air respirator
Impervious gloves
Paint suit

Tack rag
#600 grit paper
Cutoff tool
Masking tape
Tape measure
Wax and grease remover

Introduction

Tri-coats are a more glamorous paint than traditional colors because of the effect light produces on the color finish. They are also more difficult to spray, making repair more costly. Because materials are very expensive, it is critical that tri-coats be sprayed properly.

Vehicle Description

Year _____ Make _____ Model _____

VIN _____

Procedure

Task Completed

1. Cut out a 3-foot × 3-foot piece of sheet metal. Measure carefully. The sheet metal should be flat. If you are using a test panel instead, skip step 1. ☐

2. Locate a test panel, and wash it with soap and water. ☐

3. Clean the sheet metal or test panel with wax and grease remover. Scuff sand with #600 grit paper. Remove the sanding residue by blowing off, clean it with wax and grease remover one last time, and then tack the panel with a tack cloth. ☐

   a. What is the purpose of and procedure for using wax and grease remover? _____

   _____

   b. Why should you use #600 instead of #180 grit paper? _____

   _____

   c. What is the reason for using a tack cloth? _____

4. For the cutout piece of metal, mix and spray one coat of self-etching primer. Allow the proper flash time. ☐

   a. What is the purpose of self-etching primer? _____

5. For the test panel, mix and apply one coat of white urethane sealer. Allow the proper flash time. ☐

   a. What is the reason for the urethane sealer? _____

   _____

6. Mix the basecoat according to the manufacturer's recommendations. Use a tack cloth over the panel or metal piece first, and then spray one coat of white basecoat over the entire panel. Allow the proper flash time. ☐

7. For the test panel, divide the panel into three sections equally and follow the same steps as close as possible to the metal cutout piece as far as masking off each section to paint. ☐

8. For the metal cutout piece, measure at the top and bottom, 1 foot from the left side. Stretch the fine line tape vertically to the mark lines. Mask off the left third of the panel, from the fine line tape to the left edge, with masking tape and paper. ☐

9. Spray another coat of basecoat over the uncovered two-thirds of the panel. Allow the proper flash time. ☐

**Task
Completed**

10. Measure at the top and bottom, 2 feet from the left edge. Stretch the fine line tape to the vertical mark lines. Now mask off the left two-thirds of the panel. ☐

11. Spray a coat of base over the remaining third of the panel. Remove all the masking paper. Allow the proper flash time. ☐

    a.   Is there any difference in the color on the three areas sprayed? If so, explain. _____

    _____

12. Mix up the pearlcoat or tri-coat. Spray one coat over the entire panel. Allow the proper flash time. ☐

    a.   Is there a different procedure for spraying pearl? If so, explain. _____

    _____

13. For the test panel, divide the painted area into a top section, middle section, and a lower section. Then follow the same steps as close as possible to the following metal cutout steps. ☐

14. Measure up from the right and left sides to 1 foot from the bottom. Stretch the fine line tape to the mark lines. Mask off the lower one-third of the panel. ☐

15. Spray a coat of pearl over the exposed two-thirds of the panel. Allow the proper flash time. ☐

16. Measure up at the left and right sides, 2 feet from the bottom of the panel. Stretch the fine line tape to the mark lines. Mask off the lower two-thirds of the panel. ☐

17. Spray a coat of pearl over the remaining third of the panel. Remove all of the masking paper and tape. Allow the proper flash time. ☐

18. Take a tack cloth to the entire panel before applying the clearcoat. ☐

19. Mix the clearcoat. Spray two wet coats of clear over the entire panel. Allow the proper flash times between coats. ☐

    a.   What is the flash time with clearcoat? _____

    b.   What is the cure time for the clear? _____

20. Once the panels are fully cured, label the back of the panel with the number of basecoats and pearlcoats that were applied to each section. Cut the panel into nine squares, using the tape lines as guides. ☐

21. Hold each panel to the vehicle to determine which combination best matches. ☐

    a.   If none match closely, what could be the problem? _____

    _____

    b.   How would you correct this problem? _____

    _____

INSTRUCTOR'S COMMENTS _____

_____

_____

Name _____ Date _____

Tri-Coat Finish

Objective
Upon completion of this activity sheet, the student should be able to spray any panel with tri-coat.

ASE Education Foundation Task Correlation

IV.B.3	Inspect and identify type of finish, surface condition, and film thickness; develop and document a plan for refinishing using a total product system. **(HP-G)**
IV.B.5	Dry or wet sand areas to be refinished. **(HP-I)**
IV.B.6	Featheredge areas to be refinished. **(HP-I)**
IV.B.11	Identify a complimentary color or shade of undercoat to improve coverage. **(HP-G)**
IV.B.16	Remove dust from area to be refinished, including cracks or moldings of adjacent areas. **(HP-I)**
IV.B.17	Clean area to be refinished using a final cleaning solution. **(HP-I)**
IV.B.18	Remove, with a tack rag, any dust or lint particles from the area to be refinished. **(HP-I)**
IV.B.19	Apply suitable primer-sealer to the area being refinished. **(HP-I)**
IV.B.20	Scuff sand to remove nibs or imperfections from a sealer. **(HP-I)**
IV.C.1	Inspect, clean, and determine condition of spray guns and related equipment (air hoses, regulators, air lines, air source, and spray environment). **(HP-I)**
IV.C.2	Select spray gun setup (fluid needle, nozzle, and cap) for product being applied. **(HP-I)**
IV.C.3	Test and adjust spray gun using fluid, air, and pattern control valves. **(HP-I)**
IV.C.4	Demonstrate an understanding of the operation of pressure spray equipment. **(HP-G)**
IV.D.1	Identify color code by manufacturer's vehicle information label. **(HP-I)**
IV.D.2	Shake, stir, reduce, catalyze/activate, and strain refinish materials. **(HP-I)**
IV.D.3	Apply finish using appropriate spray techniques (gun arc, angle, distance, travel speed, and spray pattern overlap) for the finish being applied. **(HP-I)**
IV.D.4	Apply selected product on test or let-down panel; check for color match. **(HP-I)**
IV.D.11	Apply multistage coats for panel blending and overall refinishing. **(HP-G)**
IV.E.2	Identify a dry spray appearance in the paint surface; correct the cause(s) and the condition. **(HP-I)**
IV.E.6	Identify orange peel; correct the cause(s) and the condition. **(HP-I)**
IV.E.10	Identify sanding marks or sand scratch swelling; correct the cause(s) and the condition. **(HP-I)**
IV.E.15	Identify poor adhesion; correct the cause(s) and the condition. **(HP-G)**
IV.E.18	Identify dirt or dust in the paint surface; correct the cause(s) and the condition. **(HP-I)**
IV.E.22	Identify die-back conditions (dulling of the paint film showing haziness); correct the cause(s) and the condition. **(HP-G)**

We Support

ASE | Education Foundation

Tools and Equipment

Any panel
DA sander
Basecoat
Pearlcoat or tri-coat
Clearcoat
HVLP, RP, or any EPA-compliant paint gun
PPS cup, lids and liners (preferred), or traditional mixing cups
Wax and grease remover
#600 grit paper
Appropriately sized sanding blocks
Epoxy primer or urethane sealer

Safety Equipment

Safety glasses or goggles
Supplied air respirator
Impervious gloves
Paint suit

Introduction

Tri-coats are more glamorous than traditional colors because of the effect light has on the color finish. They are more difficult to spray, making them more costly to repair. Because materials are very expensive, it is critical that tri-coats be sprayed properly.

Vehicle Description

Year _____ Make _____ Model _____

VIN _____

Procedure

Task Completed

1. Wash the panel with the appropriate soap and water.   ☐

   a. What can happen if car wash soap is not used? _____

   _____

   b. Can dishwashing soap be used? Why or why not? _____

   _____

2. Degrease the panel.   ☐

   a. How is a vehicle degreased? _____

   b. What can happen if the vehicle or panel is not degreased? _____

   _____

3. Sand the panel with #600 grit wet sandpaper by hand or #600 grit dry sandpaper on a   ☐
   DA sander. Remove the sanding residue by blowing and tacking off.

   a. Can the panel be sanded with #320 grit by hand instead? If not, explain. _____

   _____

4. Spray on a coat of epoxy primer or urethane sealer. Allow the proper flash time.   ☐

   a. What is the difference between using the epoxy primer and urethane sealer? _____

   _____

5. Mix the basecoat. Tack off the panel with a tack cloth. Spray three medium wet coats of   ☐
   basecoat allowing proper flash times between each coat.

   a. What can happen to the color if sprayed too wet? _____

   b. What can happen to the color if sprayed too dry? _____

   c. What is the proper flash time between coats? _____

**Task
Completed**

6. Now take the paint gun apart and clean it (and the PPS cup if using one) thoroughly.   ☐

    a.   Is your shop equipped with a gun cleaner? _____

    b.   If so, can the paint in the gun be dumped into the gun cleaner? Why or why not? _

       _____

    c.   Why is it important to take the gun apart and clean it thoroughly? _____

       _____

7. Mix the pearlcoat or tri-coat. Pour it into the cup, and load it onto the spray gun.   ☐

    a.   What precautions should be taken when pouring paint into the gun? _____

       _____

8. Tack off the basecoat with a tack cloth.   ☐

9. Spray two or three coats of pearl depending on how many coats it will require for proper   ☐
   color match. Allow proper flash times in between each coat.

    a.   How do you know how many coats of pearl it will require? _____

       _____

10. Check the pearlcoat for signs of mottling. If needed, spray a third coat at an angle to the   ☐
    second coat to correct the mottling. Allow proper flash time between coats.

    a.   What is mottling? _____

    b.   How is mottling corrected? _____

11. Now take the paint gun apart and clean it (and the PPS cup if using one) thoroughly.   ☐

12. Tack off the pearlcoat with a tack cloth.   ☐

13. Mix the clearcoat. Spray on two wet coats. Allow proper flash times. Allow the proper   ☐
    cure time.

    a.   What is the most dangerous ingredient in clear? _____

    b.   What can happen if the clear is sprayed too dry? _____

    c.   How would you correct this? _____

    d.   What can happen if the clear is sprayed too wet? _____

    e.   How would you correct this? _____

INSTRUCTOR'S COMMENTS _____

_____

_____

Name _____ Date _____

Tri-Coat Spot Repair

Objective

Upon completion of this activity sheet, the student should be able to make repairs in a tri-coat.

ASE Education Foundation Task Correlation

IV.B.3	Inspect and identify type of finish, surface condition, and film thickness; develop and document a plan for refinishing using a total product system. **(HP-G)**
IV.B.5	Dry or wet sand areas to be refinished. **(HP-I)**
IV.B.6	Featheredge areas to be refinished. **(HP-I)**
IV.B.11	Identify a complimentary color or shade of undercoat to improve coverage. **(HP-G)**
IV.B.12	Apply primer onto surface of repaired area. **(HP-I)**
IV.B.13	Apply two-component finishing filler to minor surface imperfections. **(HP-I)**
IV.B.14	Block sand area to which primer-surfacer has been applied. **(HP-I)**
IV.B.15	Dry sand area to which finishing filler has been applied. **(HP-I)**
IV.B.16	Remove dust from area to be refinished, including cracks or moldings of adjacent areas. **(HP-I)**
IV.B.17	Clean area to be refinished using a final cleaning solution. **(HP-I)**
IV.B.18	Remove, with a tack rag, any dust or lint particles from the area to be refinished. **(HP-I)**
IV.B.19	Apply suitable primer-sealer to the area being refinished. **(HP-I)**
IV.B.20	Scuff sand to remove nibs or imperfections from a sealer. **(HP-I)**
IV.D.1	Identify color code by manufacturer's vehicle information label. **(HP-I)**
IV.D.2	Shake, stir, reduce, catalyze/activate, and strain refinish materials. **(HP-I)**
IV.D.3	Apply finish using appropriate spray techniques (gun arc, angle, distance, travel speed, and spray pattern overlap) for the finish being applied. **(HP-I)**
IV.D.6	Apply basecoat/clearcoat for panel blending and panel refinishing. **(HP-I)**
IV.D.8	Remove nibs or imperfections from basecoat. **(HP-I)**
IV.D.11	Apply multistage coats for panel blending and overall refinishing. **(HP-G)**
IV.E.2	Identify a dry spray appearance in the paint surface; correct the cause(s) and the condition. **(HP-I)**
IV.E.5	Identify clouding (mottling and streaking in metallic finishes); correct the cause(s) and the condition. **(HP-I)**

We Support

ASE | **Education Foundation**

Tools and Equipment

Tri-coat refinished panel
Basecoat
Pearlcoat
Clearcoat

Safety Equipment

Safety glasses or goggles
Supplied air respirator
Impervious gloves
Paint suit

Self-etching primer
#600–800 grit wet paper
HVLP, RP, or an EPA-compliant paint gun
#180, #320, #400, and #1000 grit paper
Appropriate sanding blocks
Wax and grease remover
Urethane primer

Introduction

Tri-coats are trickier and more expensive to repair than other paint finishes. Proper procedure is required when making these repairs.

Vehicle Description

Year _____ Make _____ Model _____

VIN _____

Procedure

Task Completed

1. Make a 4-inch scratch in the middle of the refinished tri-coat panel. Clean the panel with wax and grease remover. ☐

    a.   What is the procedure for using wax and grease remover? _____

    _____

2. Repair the scratch by block sanding with #180 grit paper followed by #320 grit for featheredging. Remove sanding residue and tack. ☐

    a.   What will happen if the #320 grit is not used? _____

    _____

3. Mix the self-etching primer if necessary. Otherwise, spray on one coat of self-etching aerosol primer and two coats of urethane primer. ☐

    a.   What is the purpose of self-etching primer? _____

    _____

    b.   What is the purpose of urethane primer? _____

    _____

4. Apply guide coat over the panel. ☐

    a.   What is a guide coat used for? _____

5. Block sand the primer with dry #320 and then with #400 grit wet sandpaper with a block. Finish with wet #600 grit. Remove all traces of the guide coat. ☐

    a.   Is wet sanding necessary? _____

    _____

6. Sand the rest of the hood with #1000 grit paper. ☐

    a.   Will #600 grit paper leave scratches? _____

    b.   If not, why use #1000 grit? _____

7. Mix up the basecoat per the manufacturer's recommendations. Apply one coat to cover the primer only. The next coat should go out 2 inches farther than the first. The third coat should be sprayed 2 inches past the second coat and cover the first two coats. Clean your gun while the panel is flashing. ☐

8. Mix the pearl according to instructions. Tack off the panel. Spray the first coat of pearl to cover the base only. Finish about 1 inch from the edge of the first pearlcoat. Allow the proper flash time.   ☐

9. Check for mottling. Spray on the second coat. Allow the proper flash time.   ☐

   a.   What is mottling? _____

       _____

   b.   How is mottling corrected? _____

       _____

10. Mix the clearcoat. Spray two wet coats over the entire panel. Allow sufficient time to dry.   ☐

11. Once cured, check color match and see if the blend is noticeable. If it is, repeat these job sheet steps until it is not noticeable.   ☐

INSTRUCTOR'S COMMENTS _____

_____

_____

Name _____ Date _____

Color Wheel

Objective

Upon completion of this activity sheet, the student should have an understanding of the color wheel.

ASE Education Foundation Task Correlation

IV.D.12   Identify and mix paint using a formula. **(HP-I)**

IV.D.13   Identify poor hiding colors; determine necessary action. **(HP-G)**

IV.D.14   Tint color using formula to achieve a blendable match. **(HP-I)**

IV.D.15   Identify alternative color formula to achieve a blendable match. **(HP-I)**

We Support

Education Foundation

Tools and Equipment

Pure blue, red, and yellow colors
White paper
Mixing vials

Safety Equipment

Safety glasses or goggles
Safety gloves
Approved respirator

Introduction

Vehicle buyers can choose from hundreds of colors. To repaint today's vehicles requires a working knowledge of the color spectrum. This exercise shows how the color wheel works.

Procedure

Task Completed

1. Construct the color wheel by placing 1-inch drops of paint on a sheet of white paper. Mix one part yellow with one part blue to make green. ☐

2. The color wheel will be arranged like numbers on a clock. Place a drop of red at 12 o'clock, blue at 3 o'clock, green at 6 o'clock, and yellow at 9 o'clock. ☐

3. Mix one part red with one part blue. Place the resulting color between clock numbers one and two on the color wheel. Mix one part blue and one part green. Place this new color between clock numbers four and five. Mix one part yellow and one part green, and place it between clock numbers seven and eight. Mix one part yellow with one part red. Position the resulting color between clock numbers 10 and 11. ☐

4. Mix two parts red and one part blue. Place it at clock number one. Then mix one part red with two parts blue, and place it at clock number two. Next, mix two parts blue with one part green. Place the new color at clock number four. Mix one part blue with two parts green, placing it at clock number five. ☐

5. Now for the other side of the color wheel. Mix two parts green with one part yellow. Place the resulting color at clock number seven. Then mix one part green with two parts yellow, placing it at clock number eight. For the following mix, blend two parts yellow with one part red, and position it at clock number 10. Finally, mix one part yellow with two parts red, and place it at clock number 11. ☐

6. Mix the color at clock number 11 with the color at clock number five. ☐

   a. What is the result? _____

Task Completed

7. Mix the color between clock numbers seven and eight with blue. ☐

   a. What is the result? _____

   _____

8. Mix the color at clock number eight with green. ☐

   a. What is the result? _____

   _____

9. Mix the color at clock number two with yellow. ☐

   a. What is the result?_____

   _____

10. Mix the color at clock number five with blue. ☐

   a. What is the result? _____

   _____

INSTRUCTOR'S COMMENTS _____

_____

_____

Name _____ Date _____

Gun Adjustments

Objective

Upon completion of this activity sheet, the student should be able to show how gun adjustments can change the color of a finish.

ASE Education Foundation Task Correlation

IV.B.5	Dry or wet sand areas to be refinished. **(HP-I)**
IV.B.6	Featheredge areas to be refinished. **(HP-I)**
IV.B.11	Identify a complimentary color or shade of undercoat to improve coverage. **(HP-G)**
IV.B.17	Clean area to be refinished using a final cleaning solution. **(HP-I)**
IV.B.18	Remove, with a tack rag, any dust or lint particles from the area to be refinished. **(HP-I)**
IV.B.19	Apply suitable primer-sealer to the area being refinished. **(HP-I)**
IV.B.20	Scuff sand to remove nibs or imperfections from a sealer. **(HP-I)**
IV.C.1	Inspect, clean, and determine condition of spray guns and related equipment (air hoses, regulators, air lines, air source, and spray environment). **(HP-I)**
IV.C.2	Select spray gun setup (fluid needle, nozzle, and cap) for product being applied. **(HP-I)**
IV.C.3	Test and adjust spray gun using fluid, air, and pattern control valves. **(HP-I)**
IV.C.4	Demonstrate an understanding of the operation of pressure spray equipment. **(HP-G)**

We Support

ASE | **Education Foundation**

Tools and Equipment

Three pieces of 1-foot × 1-foot sheet metal or three spray out cards
Wax and grease remover
Self-etching primer
Silver or gold metallic basecoat
Paint suit
HVLP spray gun
#400 grit paper
Tape measure (needed for marking sheet metal cut locations only)

Safety Equipment

Safety glasses or goggles
Paint respirator
Impervious gloves

Introduction

Even the most inexpensive paint guns can produce a quality paint job if the person behind the trigger knows how to effectively operate the tool. However, inexpensive paint guns just do not have the longevity of the quality guns.

Procedure

Task Completed

1. Cut out three pieces of 1-foot × 1-foot sheet metal. Measure carefully. The panels should be flat. Remove any arch by hammering. ☐

2. Or, place three different spray out cards on a holder to test spray with. This is much easier than step 1. ☐

Task
Completed

3. Clean the sheets or spray out cards with a wax and grease remover. Blow and tack off. ☐

   a.   What is the process for using the wax and grease remover? _____

     _____

   b.   Can the metal panels be hand-scuffed with #320 grit for better adhesion? _____

4. If using cutout sheet metal pieces, spray two coats of self-etching primer. ☐

5. If using the spray out cards and they are not pre-sealed, mix and apply the appropriate shade of sealer. ☐

6. Sand any dirt nibs or imperfections from the sealer. ☐

   a.   What grit should you use for removing any trash or dirt nibs? _____

7. For this demonstration to work, you need to use a consistent spray technique. Mix the paint per label instructions, and pour the paint into your gun. Spray the gun by adjusting the fan and setting the air pressure and fluid. ☐

   a.   What brand is the basecoat you are using? _____

8. Label the backside of the panels as follows: ☐

Normal

10 lb. less air pressure than normal

10 lb. more air pressure than normal

9. Spray the piece labeled Normal. Set this panel aside. ☐

   a.   What was the air pressure setting? _____

10. Adjust the gun to 10 lb. less air pressure. Obtain the proper panel. Spray this panel, remembering to use a consistent technique. ☐

   a.   What is the air pressure setting now? _____

11. Adjust the gun to 10 lb. more than the normal air pressure. Obtain the appropriate panel and spray. ☐

   a.   What is the air pressure setting now? _____

12. After the proper drying time, mix and apply clearcoat to all the test cards. There are also many aerosol spray cans now that will simulate clearcoat appearances. These also save time and money for the shop and the painter. Now compare the modified panels to the normal panel. ☐

   a.   How did the 10 lb. more air pressure panel's color match compared to the Normal panel?

     _____

   b.   How did the 10 lb. less air pressure panel's color match compared to the Normal panel?

     _____

INSTRUCTOR'S COMMENTS _____

_____

_____

Name _____ Date _____

Custom Painting Flames

Objective
Upon completion of this activity sheet, the student should be able to gain experience in custom painting.

ASE Education Foundation Task Correlation

IV.B.4	Remove paint finish as needed. **(HP-I)**
IV.B.5	Dry or wet sand areas to be refinished. **(HP-I)**
IV.B.6	Featheredge areas to be refinished. **(HP-I)**
IV.B.16	Remove dust from area to be refinished, including cracks or moldings of adjacent areas. **(HP-I)**
IV.B.17	Clean area to be refinished using a final cleaning solution. **(HP-I)**
IV.B.18	Remove, with a tack rag, any dust or lint particles from the area to be refinished. **(HP-I)**
IV.B.19	Apply suitable primer-sealer to the area being refinished. **(HP-I)**
IV.B.20	Scuff sand to remove nibs or imperfections from a sealer. **(HP-I)**
IV.D.2	Shake, stir, reduce, catalyze/activate, and strain refinish materials. **(HP-I)**
IV.D.3	Apply finish using appropriate spray techniques (gun arc, angle, distance, travel speed, and spray pattern overlap) for the finish being applied. **(HP-I)**
IV.D.6	Apply basecoat/clearcoat for panel blending and panel refinishing. **(HP-I)**
IV.D.7	Apply basecoat/clearcoat for overall refinishing. **(HP-G)**
IV.D.8	Remove nibs or imperfections from basecoat. **(HP-I)**
IV.E.2	Identify a dry spray appearance in the paint surface; correct the cause(s) and the condition. **(HP-I)**
IV.E.6	Identify orange peel; correct the cause(s) and the condition. **(HP-I)**
IV.E.10	Identify sanding marks or sand scratch swelling; correct the cause(s) and the condition. **(HP-I)**
IV.E.13	Identify tape tracking; correct the cause(s) and the condition. **(HP-G)**
IV.E.18	Identify dirt or dust in the paint surface; correct the cause(s) and the condition. **(HP-I)**
IV.E.26	Identify buffing-related imperfections (swirl marks, wheel burns); correct the condition. **(HP-I)**
IV.F.2	Sand, buff, and polish fresh or existing finish to remove defects as required. **(HP-I)**

We Support

Education Foundation

Tools and Equipment
Any hood
#400, #600, #1000, and #2000 grit paper
Masking paper
Appropriate paper

Safety Equipment
Safety glasses or goggles
Supplied air respirator
Impervious gloves
Paint suit

Sealer
Wax and grease remover
Three HVLP paint guns
Black, red, orange, and yellow basecoat

Introduction

Although many people can learn to paint professionally, few can custom paint. Painting is a skill that can be developed, but the best custom painters are born with an artistic gift. These people can command high salaries because of the demand for their work.

Procedure

Task Completed

1. Clean the hood with wax and grease remover.   ☐

    a.   What is the process and reason for using wax and grease remover? _____

    _____

2. Sand with #400 grit paper by hand or #240 grit paper on a DA sander. Remove the   ☐
    sanding residue by blowing and tacking.

    a.   Can the panel be sanded with #240 grit by hand? _____

    b.   Why or why not? _____

    _____

3. Spray one coat of sealer or epoxy primer on the hood. Allow the proper flash time.   ☐

    a.   What is the purpose of sealer? _____

    _____

    b.   What is the purpose of epoxy primer? _____

    _____

4. Nib sand as needed.   ☐

    a.   What grits are needed to nib sand? _____

5. Spray on three coats of black basecoat. Allow proper flash time between coats.   ☐

    a.   What is the flash time? _____

6. Mix and spray clearcoat according to the paint manufacturer's instructions.   ☐

    a.   What is the cure time of the clear you are using before it can be sanded and remasked
    for the next steps? _____

    _____

7. Use fine line ⅛-inch masking tape to lay out a flame design. Mask off areas that are to   ☐
    remain black with paper and tape. Sand the areas to be painted with #600 grit. Be sure
    to sand at the edges of the tape. Sand the clearcoat to a dull finish and no shiny left. Do
    not cut through it.

    a.   Why is #600 grit used instead of #400 grit? _____

    _____

8. Load the red, orange, and yellow base into separate paint guns. The idea is to spray the   ☐
    colors at the same time so the edges melt together. Spray yellow base on the front third
    of the design. Immediately spray orange on the next third. Then spray the red on the
    last third.

9. Spray other coats as needed in the same manner. When the black is totally covered, you ☐
can stop. Spray one coat of clear over the three coats. Make sure this is a full wet coat
of clear.

   a.   What can happen to the finish if the clear is sprayed too dry? _____

   _____

   b.   How would you correct this problem? _____

   _____

10. Remove the tape as soon as the clear is dust free. Allow the coat to dry per the ☐
manufacturer's recommendations.

    a.   How soon can you remove the tape? _____

    b.   What is the recommended curing time for the clear you sprayed? _____

NOTE: The next steps to wet sand and buff the finish smooth are not related to or in this
chapter. They are listed here in a small form so that you may understand how to really finish
a custom-painted panel once clearcoat is finally done. These are **NOT** the full steps! Please
refer to Job Sheet 30-1 in the next chapter.

11. Wet sand all the clearcoat with #1000 grit paper. This sanding is intended to remove ☐
the tape edge. Remove all the orange peel. Be very careful not to sand through the
clearcoat and into the base color.

    a.   If you sand into the base, how will you correct the problem? _____

    _____

12. Spray two coats of clear over the entire hood. Allow the proper drying time, then wet ☐
sand with #1500 grit, and then trizact wet sand with a DA sander and #3000 grit until all
sanded areas are a dull satin finish.

13. Now buff and polish with the appropriate pads, compounds, and polishes until ☐
completely shiny and no swirls are left.

INSTRUCTOR'S COMMENTS _____

_____

_____

Review Questions

Name _____ Date _____ Instructor Review _____

1. The primary colors are _____, _____, _____, and _____.

2. The three dimensions of color are _____ _____, _____, and _____.

3. Technician A says that tinting is only done for panel painting. Technician B says to always tint within the original mixing formula. Who is correct?
   A. Technician A
   B. Technician B
   C. Both Technician A and Technician B
   D. Neither Technician A nor Technician B

4. Metamerism results from different light sources.
   A. True
   B. False

5. Technician A adds white to increase the value. Technician B adds black to decrease the value. Who is correct?
   A. Technician A
   B. Technician B
   C. Both Technician A and Technician B
   D. Neither Technician A nor Technician B

6. When testing a color, Technician A adds tints as needed to change the color. Technician B adds tints only from the formula. Who is correct?
   A. Technician A
   B. Technician B
   C. Both Technician A and Technician B
   D. Neither Technician A nor Technician B

7. A slower drying solvent will make a metallic color darker.
   A. True
   B. False

8. When checking a metallic color for match, Technician A looks at the color head on. Technician B looks at the color from the side. Who is correct?
   A. Technician A
   B. Technician B
   C. Both Technician A and Technician B
   D. Neither Technician A nor Technician B

9. All of the following will make a metallic color darker except:
   A. opening the fluid valve.
   B. decreasing the gun distance.
   C. increasing the fan width.
   D. slowing down the stroke.

10. Technician A says that color flip-flop can be corrected by adding white tint. Technician B says the first approach to correcting this is to adjust your spraying technique. Who is correct?
    A. Technician A
    B. Technician B
    C. Both Technician A and Technician B
    D. Neither Technician A nor Technician B

Paint Problems and Final Detailing

Name _____ Date _____ Instructor Review _____

"As You Paint" Problems

These are common problems that occur while painting. You must correct these problems while you are in the booth the best way you know how. Explain how to solve each problem in a paragraph.

1. Dirt in the basecoat

_____

_____

_____

2. Dirt in the clearcoat

_____

_____

_____

3. Fish-eyes

_____

_____

_____

4. Lifting

_____

_____

_____

5. Sand scratch swelling

    _____

    _____

    _____

6. Mottling

    _____

    _____

    _____

7. Air hose dragged over wet paint

    _____

    _____

    _____

8. Tack rag gouges wet paint

    _____

    _____

    _____

9. Paint suit touches wet paint

    _____

    _____

    _____

10. Runs

    _____

    _____

    _____

Name _____ Date _____ Instructor Review _____

What Is the Problem?

1.

Name of problem: _____

How can you fix this? _____

_____

_____

_____

_____

_____

_____

2.

Name of problem: _____

How can you fix this? _____

_____

_____

_____

_____

_____

_____

3.

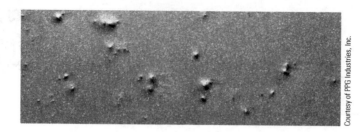

Courtesy of PPG Industries, Inc.

Name of problem: _____

How can you fix this? _____

_____

_____

_____

_____

_____

_____

4.

Courtesy of PPG Industries, Inc.

Name of problem: _____

How can you fix this? _____

_____

_____

_____

_____

_____

_____

5.

Courtesy of PPG Industries, Inc.

Name of problem: _____

How can you fix this? _____

_____

_____

_____

_____

_____

_____

6.

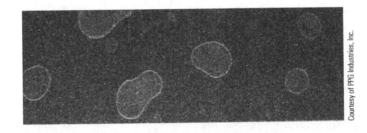

Courtesy of PPG Industries, Inc.

Name of problem: _____

How can you fix this? _____

_____

_____

_____

_____

_____

_____

7.

Courtesy of PPG Industries, Inc.

Name of problem: _____

How can you fix this? _____

_____

_____

_____

_____

_____

_____

8.

Courtesy of PPG Industries, Inc.

Name of problem: _____

How can you fix this? _____

_____

_____

_____

_____

_____

_____

9.

Courtesy of PPG Industries, Inc.

Name of problem: _____

How can you fix this? _____

_____

_____

_____

_____

_____

_____

Name _____ Date _____

Wet Sanding and Buffing Fresh Urethane Paint

Objective
Upon completion of this activity sheet, the student should be able to wet sand and buff out dirt and minor orange peel from a fresh urethane basecoat/clearcoat paint job.

ASE Education Foundation Task Correlation

IV.E.6	Identify orange peel; correct the cause(s) and the condition. **(HP-I)**
IV.E.14	Identify low gloss condition; correct the cause(s) and the condition. **(HP-G)**
IV.E.18	Identify dirt or dust in the paint surface; correct the cause(s) and the condition. **(HP-I)**
IV.E.26	Identify buffing-related imperfections (swirl marks, wheel burns); correct the condition. **(HP-I)**
IV.E.27	Identify pigment flotation (color change through film build); correct the cause(s) and the condition. **(HP-G)**

We Support

Education Foundation

Tools and Equipment
Any clearcoated panel
Buffer
Wool pad
Foam pad
Water squeeze bottle
#1000, #1500, #3000 grit sandpaper
Wooden paint mix stick or plastic sanding stick
Sand block or pad
Squeegee
Appropriate buffing/rubbing compounds and machine polish
Hand glaze

Safety Equipment
Safety glasses or goggles

Introduction
Even in today's billion dollar auto manufacturer paint facilities, dirt cannot always be eliminated. In modern automotive repair shops with the best booths and practices, dirt still finds a way to get into paint. Proper wet sanding and buffing can save the almost perfect paint job. Improper procedure can ruin a paint job, causing significant waste in terms of time and material.

Task Completed

Procedure

1. Examine the paint surface. Note the location of the dirt and excess orange peel. ☐

    a. If you are the person who painted the vehicle, what is the reason for the dirt? _____

    _____

    b. What is the reason for the excessive orange peel? _____

    _____

c.   In the painting process, how could the orange peel have been eliminated? _____

_____

2.   For dirt, allow #1000 and #1500 grit wet sandpaper to soak in water for 10 minutes.   ☐

a.   What is the reason for soaking the paper? _____

b.   What can happen if the paper is not soaked? _____

3.   For large pieces of dirt or debris, cut 1 inch of the paint stick and wrap the #1000 grit   ☐
paper around it. For smaller pieces of dirt or debris, cut #1500 grit paper and wrap
around the sanding stick. Sand the dirt in a diagonal motion. Try not to sand the adjacent
paint. Sand until the dirt is level with the surrounding area.

a.   Why should you sand diagonally? _____

_____

b.   How do you determine whether the piece actually can be wet sanded instead of just
repainted? _____

_____

4.   For excessive orange peel, place #1500–2000 grit paper on a DA and sand out the   ☐
excessive orange peel. This is much faster than hand sanding with wet sand paper.
When the texture is smooth, sand again with a DA, this time using #3000 grit paper wet
with a squirt bottle. When the finish is uniformly dull, the area is ready to be buffed.

a.   If there are any shiny spots or hazy spots left, what does this indicate? _____

_____

5.   After your instructor demonstrates the proper technique, clean the buffer pad with the   ☐
proper tool.

a.   What kind of buffing pad are you going to use? _____

b.   What is the reason for starting with this pad? _____

_____

c.   What is the proper tool to clean the pad? _____

6.   Apply a 6-inch ribbon of rubbing compound or equivalent to the surface. Working in one   ☐
small area at a time, move the buffer back and forth slowly. If the buffer speed is
adjustable, set the rpm at 1,000 or slow to medium. Be very careful around edges and
body lines. The buffer will grab onto these areas and cut through the paint way more
quickly than you will see it. Do not apply excessive pressure to the panel with the buffer.

a.   What kind of compound are you using? _____

b.   Why will the paint burn off the edges more quickly than the rest of the panel? _____

_____

c.   What can happen to the panel if excessive pressure is used? _____

_____

7. Sanded areas should brighten after buffing. You will need to buff until you remove all ☐
scratches and the surface is glossy with no hazy areas.

   a. Can the buffing pad put scratches in the paint? If so, how? _____

   _____

   b. How would you remove the scratches? _____

   _____

8. Carefully inspect the surface for more sand scratches. If you find them, continue buffing. ☐
You will notice that buffing with a wool pad will somewhat dull the paint finish. Foam
compounding pads do not do this as bad.

9. When all the scratches are eliminated, switch to a foam polishing pad and the machine ☐
glaze/polish. Apply the machine glaze/polish and buff again, holding the pad flat on the
surface.

   a. What glaze/polish are you using with the foam pad? _____

   b. Why is it important to hold the pad flat on the surface? _____

   _____

10. Once polished, wash the panel with soap and water and then dry it off. If there are still ☐
hazy spots or scratches left, repeat steps 6–9 until the surface is completely scratch
free and shiny. No hazy areas should remain.

INSTRUCTOR'S COMMENTS _____

_____

_____

Name _____ Date _____

Buffing a Faded Paint Job——Single Stage or Basecoat/Clearcoat

Objective

Upon completion of this activity sheet, the student should be able to buff a faded paint job whether it be a single-stage or a basecoat/clearcoat system.

ASE Education Foundation Task Correlation

IV.B.2	Soap and water wash entire vehicle; use appropriate cleaner to remove contaminants. **(HP-I)**
IV.B.3	Inspect and identify type of finish, surface condition, and film thickness; develop and document a plan for refinishing using a total product system. **(HP-G)**
IV.E.14	Identify low gloss condition; correct the cause(s) and the condition. **(HP-G)**
IV.E.22	Identify die-back conditions (dulling of the paint film showing haziness); correct the cause(s) and the condition. **(HP-G)**
IV.E.23	Identify chalking (oxidation); correct the cause(s) and the condition. **(HP-G)**
IV.E.26	Identify buffing-related imperfections (swirl marks, wheel burns); correct the condition. **(HP-I)**
IV.F.2	Sand, buff, and polish fresh or existing finish to remove defects as required. **(HP-I)**
IV.F.3	Clean interior, exterior, and glass. **(HP-I)**
IV.F.4	Clean body openings (door jambs and edges, etc.). **(HP-I)**
IV.F.5	Remove overspray. **(HP-I)**

We Support
ASE | Education Foundation

Tools and Equipment

Any vehicle with a faded paint job
Buffer
Wool pad
Foam pad
Pad cleaning tool
Appropriate buffing compounds and machine glaze/polishes
Hand glaze

Safety Equipment

Safety glasses or goggles

Introduction

Fewer vehicles today have a single-stage paint on them. Without the protection of a clearcoat, many of these single-stage paints fade quickly. With simple buffing, the beauty of the original paint job can be brought back to life.

Vehicle Description

Year _____ Make _____ Model _____

VIN _____

Procedure

Task Completed

1. Examine the paint. Usually the upper surfaces fade the most; therefore, the upper surfaces will need the most buffing.   ☐

    a.   Why do the upper surfaces fade more than the sides? _____

    _____

2. Wash the complete vehicle with soap and water even if you are only buffing one panel.   ☐

    a.   What is the reason for this? _____

    _____

    b.   What type of soap did you use to wash the car? _____

3. Using the wool pad or foam pad (a wool pad will not only bring a shine faster but also burn through the paint faster), buff one 2-foot × 2-foot area at a time. Start on the roof. Apply the buffing compound to the area you are about to buff.   ☐

    a.   Why do you buff one 2-foot × 2-foot area at a time? _____

    _____

    b.   Why do you start with the roof? _____

    _____

    c.   Would you use the same compound with the foam pad as you would use with a wool pad? _____

    _____

4. Work the buffing compound into the 2 foot × 2 foot area with the buffing pad before spinning the buffer.   ☐

    a.   Why should you work the buffing compound into the 2 foot × 2 foot area with the buffing pad before spinning the buffer? _____

    _____

5. Keep the buffer at a very slight angle to the surface. Apply moderate pressure to the surface as you slowly work the area back and forth. Keep the electrical cord over your shoulder.   ☐

    a.   What is the reason for keeping the cord over your shoulder? _____

    _____

    b.   Why do you keep the buffer at an angle? _____

    _____

    c.   What can happen if too much pressure is applied with the buffer? _____

    _____

6. Be careful when buffing on or around the body lines and panel edges.   ☐

    a.   What can happen if you buff directly on the body lines and edges? _____

    _____

b.   Why does the paint burn through so easily on these edges? _____

_____

c.   How can you correct burn-through? _____

_____

7. You can apply pressure as long as you have buffing compound between the pad and   ☐
the paint. When the compound is used up, do not apply pressure to the area.

a.   What can happen to the paint if there is no compound on the panel? _____

_____

b.   Feel the panel immediately after buffing; is the panel warm? _____

c.   What can happen if you stay on one spot too long? _____

_____

8. When you have finished the first area, move on to the next area of the roof and continue   ☐
buffing. Continue with the other top panels and then any side panels that may need it as
well until the complete vehicle is done. Clean the pad with the proper tool if it becomes
clogged with compound.

a.   What tool is used to clean the pad? _____

9. Do not buff any longer than it takes to bring back the shine. There is a finite amount of   ☐
paint on the vehicle.

10. If there are any dull, faded, or hazy spots still visible, repeat steps 3–7 again until no   ☐
more of these places are visible.

11. When you have finished buffing the entire vehicle, you can use the foam pad with   ☐
machine glaze/polish. Move to a new pad. It is important to hold the buffer with the
foam pad *flat.*

a.   What is the reason for holding the buffer flat to the surface at this point? _____

_____

12. Using the glaze will remove most of the swirl marks caused by the buffer. After quickly   ☐
buffing the vehicle again with the polish, you should wash the car to remove all the
compound from the openings and windows.

13. The vehicle is now ready for final inspection. Although buffing removes the oxidation   ☐
and the dead paint, it does not give the paint any protection. Apply a hand glaze, and
remove it with a soft rag.

14. If there are any swirl marks still visible, repeat steps 9–11 again until no more swirls   ☐
are visible.

INSTRUCTOR'S COMMENTS _____

_____

_____

Name _____ Date _____

Correcting Common Imperfections

Objective

Upon completion of this activity sheet, the student should be able to recognize and correct a fish-eye problem in the paint.

ASE Education Foundation Task Correlation

IV.B.2	Soap and water wash entire vehicle; use appropriate cleaner to remove contaminants. **(HP-I)**
IV.B.3	Inspect and identify type of finish, surface condition, and film thickness; develop and document a plan for refinishing using a total product system. **(HP-G)**
IV.B.4	Remove paint finish as needed. **(HP-I)**
IV.B.5	Dry or wet sand areas to be refinished. **(HP-I)**
IV.B.6	Featheredge areas to be refinished. **(HP-I)**
IV.B.11	Identify a complimentary color or shade of undercoat to improve coverage. **(HP-G)**
IV.B.12	Apply primer onto surface of repaired area. **(HP-I)**
IV.B.14	Block sand area to which primer-surfacer has been applied. **(HP-I)**
IV.B.16	Remove dust from area to be refinished, including cracks or moldings of adjacent areas. **(HP-I)**
IV.B.17	Clean area to be refinished using a final cleaning solution. **(HP-I)**
IV.B.18	Remove, with a tack rag, any dust or lint particles from the area to be refinished. **(HP-I)**
IV.D.2	Shake, stir, reduce, catalyze/activate, and strain refinish materials. **(HP-I)**
IV.D.3	Apply finish using appropriate spray techniques (gun arc, angle, distance, travel speed, and spray pattern overlap) for the finish being applied. **(HP-I)**
IV.D.6	Apply basecoat/clearcoat for panel blending and panel refinishing. **(HP-I)**
IV.D.7	Apply basecoat/clearcoat for overall refinishing. **(HP-G)**
IV.D.8	Remove nibs or imperfections from basecoat. **(HP-I)**
IV.D.16	Identify the materials equipment and preparation differences between solvent and waterborne technologies. **(HP-G)**
IV.E.3	Identify the presence of fish-eyes (crater-like openings) in the finish; correct the cause(s) and the condition. **(HP-I)**
IV.E.4	Identify lifting; correct the cause(s) and the condition. **(HP-G)**
IV.E.7	Identify overspray; correct the cause(s) and the condition. **(HP-I)**
IV.E.10	Identify sanding marks or sand scratch swelling; correct the cause(s) and the condition. **(HP-I)**
IV.E.11	Identify contour mapping/edge mapping; correct the cause(s) and the condition. **(HP-G)**
IV.F.5	Remove overspray. **(HP-I)**

We Support

ASE | Education Foundation

Tools and Equipment

Any panel
HVLP, RP, or any EPA-compliant paint gun
Basecoat
#400 and #600 grit paper
Epoxy primer
Wax and grease remover
Silicone spray
Cleaning cloths
Blowgun
Tack cloth
Masking paper
Various sized masking tape

Safety Equipment

Safety glasses or goggles
Paint respirator
Impervious gloves
Paint suit

Introduction

There are many problems that can arise when refinishing a vehicle. Fish-eyes are a common problem and an easy one to correct. The most common causes are a lack of proper solvent cleaning and silicone in the air.

Procedure

Task Completed

Fish-Eyes

1. Wash the panel with soap and water. Clean with wax and grease remover. ☐
   a. What soap did you use to wash the panel? _____
   b. How can the wrong soap cause fish-eyes? _____
   _____

2. If possible, mask the panel edges off by back taping. Wet sand the panel with #600 grit ☐
   by hand or dry sand with #600 on a DA. Blow off and tack.
   a. What will happen if you sand the panel with a lower grit instead? _____
   _____
   b. How is the problem corrected? _____
   _____

3. Mix the basecoat as per label instructions. ☐
   a. What brand of base are you using? _____

4. Load base into the gun, and spray the panel. ☐
   a. Describe in full detail what the paint looks like after spraying. _____
   _____

5. Use a heat lamp or dryer to quickly dry the panel. ☐

6. Clean the panel with wax and grease remover. Wet sand the panel with #600 grit. ☐
   Be careful not to dig into the fresh paint. Sand out all the craters.
   a. What is the purpose of using the #600 grit? _____
   _____

7. Clean the panel again with wax and grease remover. Blow it off and tack. ☐
   a. What is the purpose for using the wax and grease remover again? _____
   _____

8. Spray on another coat of base. If the fish-eyes reappear, repeat the earlier process. ☐

9. If the fish-eyes are very minor, sometimes spraying the basecoat very dry will cover the fish-eyes without repeated sanding. ☐

10. Check for any overspray that might have made it through the masking paper or tape. ☐

    a. What will you use to properly remove the overspray? _____

    _____

Lifting

11. Lifting is a problem resulting from incompatibility in the refinish materials. This can be from spraying lacquer-based products over repainted enamel. This can also happen mainly from freshly painting a panel, having a defect or some reason to need to repaint it, and it lifting when refinishing it. Wash the fender with the appropriate car wash soap. Clean with a wax and grease remover. Sand with #400 grit paper. ☐

    a. Why is the proper car wash soap important? _____

    _____

    b. How do you properly use wax and grease remover? _____

    _____

12. Mix the basecoat/clearcoat topcoats, and spray as recommended. Allow the paint to dry overnight, or run through the bake cycle. ☐

    a. Was the clearcoat a quick clear or an overall clear, and what brand? _____

    _____

13. Make a 6-inch scratch in the fender. Repair the scratch by block sanding with #320 grit. Featheredge at least 1 inch of each layer. At this point, you should have three layers of different material. ☐

14. Run a tape line halfway across the sanded area. Apply masking paper to cover up half of the repaired area. Once half of the sanded repair area is covered up, you may spray the visible bare metal with self-etching primer. ☐

    a. What is the purpose of the self-etching primer? _____

    _____

    b. What is the purpose of covering half of the repaired sanded area? _____

    _____

15. Mix and spray two coats of urethane primer. ☐

    a. What is the purpose of urethane primer? _____

    _____

16. Next, apply a guide coat. ☐

    a. What is the purpose of a guide coat? _____

    _____

17. Block sand the primer with wet #600 grit. Remove all the guide coat. Blow off and tack. ☐

18. Mix and apply a sealer over the uncovered area. ☐

Task Completed

19. Remove the masking paper and tape from the other half of the repaired area. ☐

20. Now, mix and spray on one wet coat of basecoat. There should be lifting where the paint was sanded through, covered up and not primed. ☐

    a. What does this lifting look like? _____

    _____

21. To quickly dry the paint, use a heat lamp for 10–15 minutes, or force dry in the paint booth cycle. After the base has dried, use wet #1000 grit paper to sand out smooth the heavy lift scratches. Do not sand too deeply, or you will roll the paint. Spray a very dry basecoat over the scratches. Apply three or more dry coats until the area has been fully covered and no lifting has reoccurred. Allow proper drying time between each coat. ☐

    a. What is the purpose of the dry spray? _____

    _____

22. Tack off and spray on a medium wet coat of paint. If lifting continues, repeat step 20 until no more lifting occurs. ☐

23. For extreme lifting problems that will not cap off with multiple coats of dry sprayed basecoat over a smooth sanded surface, you must sand the entire panel with #600 grit to completely correct the lifting problem. Clean with wax and grease remover; tack off and spray one coat of a water-based primer. ☐

    a. What is the water-based primer used for, and how will it correct the issue? _____

    _____

24. Allow the water-based primer to dry according to the product manufacturer's recommendations. Sand with #600 grit wet sand paper to remove the orange peel, and recoat with base. ☐

    a. After recoating, did the water-based primer correct the problem? _____

    _____

INSTRUCTOR'S COMMENTS _____

_____

_____

Review Questions

Name _____ Date _____ Instructor Review _____

1. The original finish seeping through the new topcoat color is called _____
_____.

2. Technician A says that solvent pop is caused by insufficient flash time. Technician B believes that the solvent pop is caused by improper use of an HVLP paint gun. Who is correct?
   A. Technician A
   B. Technician B
   C. Both Technician A and Technician B
   D. Neither Technician A nor Technician B

3. Any nearby textured trim should be masked off before buffing or polishing.
   A. True
   B. False

4. "Fish-eyes" are caused by _____
_____.

5. Burn-through when buffing is most likely to happen on the edges and crowns of panels.
   A. True
   B. False

6. Before buffing a newly painted urethane clearcoat, Technician A uses #1000 grit paper to remove any runs or large pieces of dirt. Technician B uses #1500 grit paper for small dirt nibs. Who is correct?
   A. Technician A
   B. Technician B
   C. Both Technician A and Technician B
   D. Neither Technician A nor Technician B

7. Technician A says paint swirl marks can be caused by dirty or worn buffing pad. Technician B says to use a wool pad instead of a foam pad to remove swirls. Who is correct?
   A. Technician A
   B. Technician B
   C. Both Technician A and Technician B
   D. Neither Technician A nor Technician B

8. To remove overspray on a windshield, Technician A uses a razor blade. Technician B uses a rag dipped in thinner. Who is correct?
   A. Technician A
   B. Technician B
   C. Both Technician A and Technician B
   D. Neither Technician A nor Technician B

9. Technician A buffs urethane clear within 24 hours of painting. Technician B buffs the urethane clear after 36 hours of painting. Who is correct?
   A. Technician A
   B. Technician B
   C. Both Technician A and Technician B
   D. Neither Technician A nor Technician B

10. The buffer should always be moving when being used.
   A. True
   B. False